Musculoskeletal Examination
of the
Spine

Making the Complex Simple

Edited by

Jeffrey A. Rihn, MD
Assistant Professor
Department of Orthopaedic Surgery
Thomas Jefferson University Hospital
The Rothman Institute
Philadelphia, Pennsylvania

Eric B. Harris, MD
Director of Orthopaedic Spine Surgery
Department of Orthopaedic Surgery
Naval Medical Center San Diego
San Diego, California

MUSCULOSKELETAL EXAMINATION
MAKING THE COMPLEX SIMPLE
SERIES

Series Editor, Steven B. Cohen, MD

D1394171

www.slackbooks.com

ISBN: 978-1-55642-996-5

Published by: SLACK Incorporated
                       6900 Grove Road
                       Thorofare, NJ 08086 USA
                       Telephone: 856-848-1000
                       Fax: 856-848-6091
                       www.slackbooks.com

Contact SLACK Incorporated for more information about other books in this field or about the availability of our books from distributors outside the United States.

Library of Congress Cataloging-in-Publication Data
Musculoskeletal examination of the spine : making the complex simple / [edited by] Jeffrey A. Rihn, Eric B. Harris.
    p. ; cm. -- (Musculoskeletal examination : making the complex simple series)
  Includes bibliographical references and index.
  ISBN 978-1-55642-996-5 (alk. paper)
  1. Spine--Imaging. 2. Spine--Diseases--Diagnosis. 3. Musculoskeletal system--Examination. I. Rihn, Jeffrey A. II. Harris, Eric B. III. Series: Musculoskeletal examination : making the complex simple series.
  [DNLM: 1. Spinal Diseases--diagnosis. 2. Physical Examination--methods. WE 725]
  RD768.M87 2011
  616.7'3075--dc22
                              2010053008

For permission to reprint material in another publication, contact SLACK Incorporated. Authorization to photocopy items for internal, personal, or academic use is granted by SLACK Incorporated provided that the appropriate fee is paid directly to Copyright Clearance Center. Prior to photocopying items, please contact the Copyright Clearance Center at 222 Rosewood Drive, Danvers, MA 01923 USA; phone: 978-750-8400; website: www.copyright.com; email: info@copyright.com

Printed in the United States of America.

Last digit is print number: 10 9   8   7   6   5   4   3   2   1

DEDICATION

This book is dedicated to my lovely wife, Theresa, and our 2 wonderful children, Maggie and Charlie. I am grateful for their unconditional love and support. —Jeffrey A. Rihn

This book is dedicated to May, Cameron, and Haley for all their love and endless patience. —Eric B. Harris

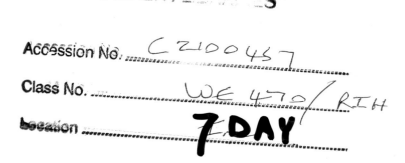

CONTENTS

ACKNOWLEDGMENTS

We would like to acknowledge and thank all of the authors for their hard work and contributions.

ABOUT THE EDITORS

Jeffrey A. Rihn, MD is an Assistant Professor of Orthopaedic Surgery at Thomas Jefferson University Hospital and The Rothman Institute. He attended the University of Pennsylvania Medical School, where he was inducted into the Alpha Omega Alpha Honors Society. He then completed a 5-year orthopedic surgery residency at the University of Pittsburgh Medical Center and a spine fellowship at Thomas Jefferson University Hospital. His practice includes the treatment of cervical, thoracic, and lumbar degenerative disorders; spinal trauma, deformity, spinal infections, and tumors of the spine. He is actively involved in numerous research projects, with a special focus on clinical outcomes research. Dr. Rihn is certified by the American Board of Orthopaedic Surgery and is an active member of the American Academy of Orthopaedic Surgery, the North American Spine Society, the American Spinal Injury Association, and the Council on the Value of Spinal Care. He has published numerous peer-reviewed studies, review articles, and book chapters on various topics of spinal care and has presented his research at numerous national and international meetings.

Eric B. Harris, MD is the Director of the Multidisciplinary Spine Center at the Naval Medical Center San Diego in San Diego, California. His specific interests include adult deformity correction, less invasive management of thoracolumbar disk degeneration, and cervical disk arthroplasty. After his residency at the Naval Medical Center San Diego, he furthered his orthopedic training with a spine surgery fellowship at The Rothman Institute and Thomas Jefferson University Hospital in Philadelphia. While there, Dr. Harris held a faculty teaching appointment at the Thomas Jefferson University School of Medicine.

Dr. Harris began his academic career at the University of Utah where he earned his bachelor of science degree in environmental earth sciences with minors in chemistry and naval science. His medical training began at the F. Edward Hebert School of Medicine at the Uniformed Services University of the Health Sciences where he obtained his medical degree.

After a basic surgery internship at the Naval Medical Center San Diego, he attended Diving Medical Officer training in Groton, Connecticut and Panama City Beach, Florida. He was then assigned to the Naval Submarine Medical Research Laboratory in Groton, CT where he completed research on Vitamin D deficiency in underway submariners as well as assisting in numerous studies on the underwater effects of sound on US Navy divers. Dr. Harris then returned to San Diego for his residency followed by his fellowship in Philadelphia.

Dr. Harris is currently stationed at the Naval Medical Center San Diego and in January of 2011 deployed to Afghanistan in support of Operation Enduring Freedom. He has produced numerous publications for peer reviewed journals, as well as book chapters, and has been invited faculty at several educational spine meetings in the United States.

CONTRIBUTING AUTHORS

Todd J. Albert, MD (Chapter 6)
Professor and Chairman
Department of Orthopedic Surgery
Professor of Neurosurgery
The Rothman Institute and
Thomas Jefferson University and Hospital
Philadelphia, Pennsylvania

R. Todd Allen, MD, PhD (Chapter 12)
Assistant Clinical Professor of Orthopedic Surgery
University of California San Diego
Department of Orthopedic Surgery
San Diego, California

D. Greg Anderson, MD (Chapter 9)
Professor, Department of Orthopedic Surgery
Thomas Jefferson University
Philadelphia, Pennsylvania

David T. Anderson, MD (Chapter 6)
Resident
Department of Orthopedic Surgery
Thomas Jefferson University Hospital
Philadelphia, Pennsylvania

Brian T. Barlow, MD (Chapter 2)
Orthopedic Resident
Naval Medical Center
San Diego, California

Mark L. Dumonski, MD (Chapter 9)
Spine and Orthopaedic Surgery
Guilford Orthopaedic and Sports Medicine Center
Greensboro, North Carolina

Matthew R. Eager, MD (Chapter 18)
SUN Orthopaedic Group, Inc
Division of Spine Surgery
Lewisburg, Pennsylvania

Ian D. Farey, FRACS (Chapter 8)
Orthpaedic Spinal Surgeon
North Shore Private Hospital
Department of Orthopaedic Surgery
St. Leonards, Sydney, New South Wales, Australia

Steven R. Garfin, MD (Chapter 12)
Chair and Professor of Orthopedic Surgery
University of California San Diego
Department of Orthopedic Surgery
San Diego, California

Greg Gebauer, MD (Chapter 16)
Advanced Orthopedic Center
Port Charlotte, Florida

Joseph P. Gjolaj, MD (Chapter 14)
Chief Resident
Department of Orthopedic Surgery
The Johns Hopkins University
Baltimore, Maryland

Kathryn H. Hanna, MD (Chapter 15)
Department of Orthopaedic Surgery
Naval Medical Center, San Diego
San Diego, California

James S. Harrop, MD, FACS (Chapter 17)
Associate Professor
Departments of Neurological and Orthopedic Surgery
Jefferson Medical College
Philadelphia, Pennsylvania

Alan S. Hilibrand, MD (Chapter 5)
Professor of Orthopedic Surgery
Director of Orthopedic Medical Education
Professor of Neurological Surgery
Jefferson Medical College
The Rothman Institute
Philadelphia, Pennsylvania

Justin B. Hohl, MD (Chapters 1 and 11)
Chief Resident
University of Pittsburgh Medical Center
Department of Orthopaedic Surgery
Pittsburgh, Pennsylvania

James D. Kang, MD (Chapter 11)
Professor of Orthopedic and Neurological Surgery
UPMC Endowed Chair in Spine Surgery
Vice Chairman, Department of Orthopedic Surgery
Director of Ferguson Laboratory for Spine Research
University of Pittsburgh School of Medicine
Pittsburgh, Pennsylvania

Joon Y. Lee, MD (Chapter 1)
Assistant Professor of Orthopedics
Division of Spine Surgery
University of Pittsburgh Medical Center
Pittsburgh, Pennsylvania

Joseph K. Lee, MD (Chapter 10)
Resident
Columbia University Medical Center
Department of Orthopedic Surgery
New York, New York

Christopher Loo, MD, PhD (Chapters 3 and 4)
Fellow
Department of Nanomedicine and Biomedical Engineering
The Methodist Hospital/Research Institute
Houston, Texas

Ryan P. Ponton, MD (Chapter 7)
Department of Orthopedics
Naval Medical Center
San Diego, California

Kris Radcliff, MD (Chapter 5)
Assistant Professor
Department of Orthopedic Surgery
Thomas Jefferson University
Rothman Institute
Philadelphia, Pennsylvania

Nelson S. Saldua, MD (Chapter 17)
Attending Orthopedic Spine Surgeon
Department of Orthopedic Surgery
Naval Medical Center
San Diego, California

Davor D. Saravanja, FRACS (Chapter 8)
Orthopaedic Spinal Surgeon
Sydney Children's Hospital
Department of Orthopaedic Surgery
Randwick, Sydney, New South Wales, Australia

Adam L. Shimer, MD (Chapter 18)
Assistant Professor of Orthopedic Surgery
Division of Spine Surgery
University of Virginia Medical Center
Charlottesville, Virginia

Harvey E. Smith, MD (Chapters 3 and 13)
Tufts University
New England Baptist Hospital
New England Orthopaedic and Spine Surgery
Boston, Massachusetts

Brian W. Su, MD (Chapter 10)
Orthopedic Spine Surgeon
Mt. Tam Spine Center
Larkspur, California

Ishaq Y. Syed, MD (Chapters 1 and 11)
Assistant Professor
Department of Orthopedic Surgery
Wake Forest University Baptist Medical Center
Winston-Salem, North Carolina

Jeffrey M. Tuman, MD (Chapter 18)
Resident Physician
Department of Orthopedic Surgery
University of Virginia Medical Center
Charlottesville, Virginia

Vidyadhar V. Upasani, MD (Chapter 12)
Resident
University of California San Diego
Department of Orthopedic Surgery
San Diego, California

Alexander R. Vaccaro, MD, PhD (Chapters 13 and 16)
Everett J. and Marion Gordon Professor of Orthopedic
    Surgery
Professor of Neurosurgery
Co-Director of the Delaware Valley Spinal Cord Injury Center
Co-Chief Spine Surgery
Co-Driector Spine Surgery
Thomas Jefferson University and the Rothman Institute
Philadelphia, Pennsylvania

W. Timothy Ward, MD (Chapter 14)
Chief of Pediatric Orthopedics
University of Pittsburgh Medical Center
Pittsburgh, Pennsylvania

Bradley K. Weiner, MD (Chapters 3 and 4)
Professor of Clinical Orthopedic Surgery
Weill Cornell Medical College
Vice Chairman of Orthopedic Surgery
Chief of Spinal Surgery
The Methodist Hospital
Business Practices Officer
Director, Spine Advanced Technology Laboratory
The Methodist Hospital Research Institute
Houston, Texas

PREFACE

The spine is a very complex entity compromised of bones, ligaments, joints, and the neural elements that it is designed to protect. Diagnosing and treating problems of the spine can be challenging. The purpose of this book was to take a very complicated topic and simplify it as much as possible in order to give the reader a general understanding of how to evaluate and manage patients with various spine-related problems. This book includes contributions from some of the world's leading experts on spinal disorders. We would like to thank all of the authors for their contributions.

FOREWORD

In regards to the musculoskeletal system, the spine can be one of the most challenging topics to grasp for those going into a healthcare profession. Because disorders of the spine are so common, however, it is important that healthcare providers have an understanding of how to perform a history and physical examination of the spine patient, how to order appropriate imaging studies and interpret them, how to make a diagnosis of the most common spinal disorders, and how to manage these disorders. This book provides an excellent overview of these topics in a comprehensible fashion. Drs. Rihn and Harris are leaders in the field of spinal surgery. Their expertise in this area, as well as the expertise of the contributing authors, is evident throughout this book, which successfully takes a very complicated subject and presents it in such a way that will be of great value to those who are trying to learn the basics about the spine. This book is well suited for medical students, allied health professionals, residents, and spine fellows who want to learn the essentials of evaluating and treating spine patients.

Todd J. Albert, MD
Professor and Chairman
Department of Orthopaedic Surgery
Thomas Jefferson University Hospital
President, The Rothman Institute
Philadelphia, Pennsylvania

INTRODUCTION

Spine-related pathology is one of the most common reasons that patients present to their physician, the most common complaint being "lower back pain". Approximately 80% of the adult population experience lower back pain at some point in their lives. For this reason, it is important for all of those practicing medicine to have some understanding of the spine. Spine patients present in many settings, including the primary care physician's office, the emergency room, and the trauma bay. There are many important components to evaluating a spine patient, including the history, physical examination, and diagnostic imaging. A thorough understanding of these components is essential to formulating an accurate diagnosis and implementing appropriate treatment. An error in diagnosis for some spinal disorders can, unfortunately, have devastating consequences.

This book is intended to provide an overview of the evaluation and treatment of the spine patient. The material is presented in a simple, understandable fashion, using figures and tables to supplement the text. A detailed description of the physical examination of the spine is provided as is an overview of the most-utilized imaging studies. Furthermore, a description of the evaluation and management of the most common spinal diseases that clinicians encounter is provided, including degenerative disease of the cervical and lumbar spine, spinal trauma, scoliosis, and spinal infection and tumor.

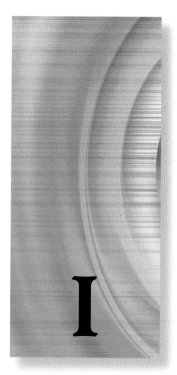

I

Physical Examination

1

PHYSICAL
EXAMINATION OF
THE CERVICAL SPINE

Justin B. Hohl, MD; Ishaq Y. Syed, MD; and Joon Y. Lee, MD

INTRODUCTION

In the current age of technology in which advanced imaging abounds, the physical examination of the cervical spine remains essential in the diagnosis of cervical pathology. Although a complete history provides the framework for a differential diagnosis, the physical examination allows the examiner to test that differential and narrow the list of possible diagnoses. Performing a thorough examination of the cervical spine identifies the extent of neurologic injury and localizes the lesion. This chapter outlines the steps of a detailed physical examination of the cervical spine, highlighting key tests and findings.

Rihn JA, Harris EB. *Musculoskeletal Examination of the Spine: Making the Complex Simple* (pp. 2-17).
© 2011 SLACK Incorporated.

INSPECTION

Like all musculoskeletal examinations, inspection is the first step in evaluating the cervical spine. Observation of a patient's posture, body positioning, demeanor, and expressions of pain or discomfort are the first visual signals that a patient gives to an examiner. Viewing the patient in the coronal and sagittal planes will give the examiner important information about spinal imbalance such as kyphotic deformities. The skin should be inspected for previous scars, café au lait spots indicative of neurofibromatosis, and signs of congenital spine anomalies such as midline tufts of hair, abnormal pigmentation, or dimpling.

Closer inspection may reveal areas of asymmetry, such as a unilateral drooping shoulder or areas of atrophy. Specifically, wasting of the shoulder girdle may indicate stenosis at C4-C5 and C5-C6 leading to chronic compression and injury to motor neurons, although the differential includes degenerative diseases of the upper motor neurons or rotator cuff pathology. Likewise, wasting of the intrinsic muscles of the hand with spasticity may implicate "myelopathy hand."[1]

Gait analysis is also an essential part of the examination, as an abnormal gait may have a neurologic etiology. Specifically, a wide-based unsteady gait is often seen in patients with cervical myelopathy, although this gait may also be seen in patients with cerebellar dysfunction.

PALPATION

Identifying areas of pain and discomfort through palpation is the next step in examining the cervical spine. Focal tenderness of bony prominences may be helpful in patients with a history of trauma. In these cases, the cervical orthosis should be loosened with care to palpate for step-offs or widening of the spinous processes that may indicate fracture or dislocation of the cervical spine. Soft-tissue palpation likewise can identify focal areas of pain and pathology in both the acute and chronic setting.

Palpating bony anatomy is also helpful in conceptualizing the complex 3-dimensional anatomy as well as preoperative planning. The carotid tubercle, also known as the Chassaignac tubercle, is the anterior tubercle of the transverse process of the C6 vertebra and may be palpable in the anterior neck, which can help in making an incision at the correct level for anterior spine surgery. Posteriorly, the spinous processes of C2 and C7 are the most prominent and again can be used as landmarks in planning incisions.

SURFACE ANATOMY

Understanding surface anatomy will help in grasping the 3-dimensional anatomy of the neck, preoperative planning, and identifying sources of pain. The vertebral bodies of C1 and C2 are located just posterior to the oropharynx. Anteriorly, the hyoid bone is located at the C3-C4 level, while the upper margin of the thyroid cartilage (the Adam's apple) is at C4-C5 and the lower margin at C5-C6. The cricoid cartilage is at C6, and the first tracheal ring is at C6-C7.[2] Regarding soft-tissue anatomy, the most prominent anterior muscle is the sternocleidomastoid, which originates on the manubrium of the sternum and the clavicle, and inserts on the mastoid process of the temporal bone of the skull.

RANGE OF MOTION

A complete evaluation of range of motion of the cervical spine includes flexion, extension, rotation, and lateral bending. Pathology in the cervical spine can result in decreased range of motion of the neck. Trauma, tumor, infection, and cervical spondylosis are all potential causes of decreased cervical range of motion. Pain or difficulty in rotating or flexing the head may be due to pathology at C1-C2, as 50% of the neck motion comes from this level.[3] Furthermore, certain positions can exacerbate pain and indicate etiologies of pathology. For example, extension may be painful in patients with stenosis or nerve root compression because extending the cervical spine decreases the space available for the cord and can compress nerve roots. Patients with a fixed rotation or tilt may have atlantoaxial rotatory subluxation or an underlying unilateral facet dislocation.

Table 1-1

MOTOR GRADING CHART

Grade	Description
5	Movement against full resistance
4	Movement against gravity with some resistance
3	Complete range of motion against gravity
2	Complete range of motion with gravity eliminated
1	Muscle flicker
0	No palpable or visible contraction

STRENGTH TESTING

Pathology involving the nerve roots and spinal cord is commonly manifest in the extremities, and every patient should have a detailed motor examination with strength grading (Table 1-1). Evaluating strength systematically will result in identification of subtle weakness patterns due to cord compression or nerve injury.

An overview of upper-extremity motor innervation is outlined in Table 1-2 and Figure 1-1. It is important to remember that in evaluating for cervical radiculopathy, cervical nerve roots exit above the corresponding pedicle, unlike the lumbar spine, where they exit below the pedicle. For example, the C5 root exits between C4 and C5, with the only exception being the C8 root, which exits above the T1 pedicle. A posterolateral C4-C5 disk herniation typically results in C5 radiculopathy.

SENSORY TESTING

Sensory testing typically consists of testing pain and light touch, which includes one function from the spinothalamic tract (pain and temperature) and one from the dorsal columns (light touch, joint position sense, and vibration). Pain can be tested with pinprick by using a clean safety pin. In patients

Table 1-2

UPPER-EXTREMITY MOTOR, SENSORY, AND REFLEX EVALUATION

Root	Motor	Sensory	Reflex
C5	Deltoid	Lateral deltoid	Biceps
C6	Biceps and wrist extension	Thumb and lateral forearm	Brachioradialis and biceps
C7	Triceps and wrist flexion	Middle finger	Triceps
C8	Finger flexion	Small finger and medial forearm	None
T1	Hand intrinsics/ interossei	Medial arm	None

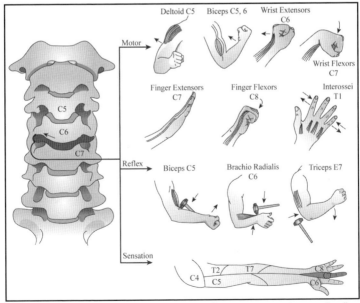

Figure 1-1. Cervical innervation of motor function, sensation, and reflex arcs in the upper extremities.

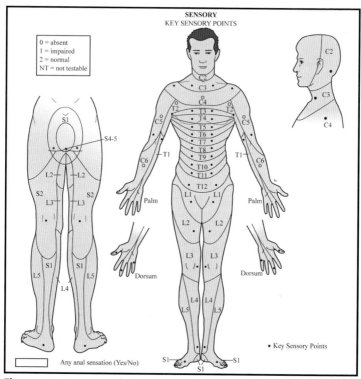

Figure 1-2. Dermatomal distribution of sensation.

suspected of having cord compression, neuropathy, or peripheral nerve entrapment, proprioception and vibration should also be tested. A dermatomal diagram is depicted in Figure 1-2, and descriptions of sensory tests were included in Table 1-2.

Vascular Testing

Focused examination of the vascular system can be helpful in certain scenarios. In patients with cervical spine trauma, awareness of the possibility of vertebral artery injury is imperative. Although it is not possible to palpate the vertebral artery, it is possible to diagnose a stroke due to vertebral artery injury on examination. Because the posterior inferior cerebellar artery arises from the vertebral artery, occlusion of this branch can result in a lateral medullary syndrome (Wallenberg syndrome), which arises from ischemia to the lateral medulla and inferior cerebellum. This is characterized by multiple possible findings, including ipsilateral cerebellar findings, dysphagia, ipsilateral facial numbness with contralateral trunk and extremity numbness, vertigo, and ipsilateral Horner syndrome.[4]

Thoracic outlet syndrome is another vascular and neurologic syndrome that should be tested for in patients with a neurologic deficit. This syndrome is caused by compression of the subclavian vessels and brachial plexus in the area between the interscalene triangle and inferior border of the axilla. Potential sources of compression include anomalous upper rib structures, trauma, or congenital soft-tissue bands. The Adson maneuver can help identify thoracic outlet syndrome by abducting, extending, and externally rotating the arm while the patient takes a deep breath and rotates his or her head toward the affected side. The test is positive if the vessels and plexus are compressed and cause a drop in palpated pulse pressure or blood pressure and reproduction of the neurologic symptoms.

Reflexes

The stretch reflex arc begins with a muscle spindle responding to stretch and transmitting a signal via an afferent peripheral nerve to the spinal cord synapse (Figure 1-3). The signal is then relayed to the efferent nerve and on to the muscle fibers, eliciting a response. Upper motor neurons descend from the brain to modulate the reflexes. Disrupting the reflex arc results in the loss of that reflex, and compression of a nerve root may lead to hyporeflexia. On the other hand, interrupting the input from upper motor neurons may lead to a relative hyperreflexia, which is commonly encountered in cervical spondylotic myelopathy.

As shown in Table 1-2, cervical reflexes include the biceps reflex for C5, brachioradialis for C6, and triceps for C7, although some overlap exists between levels. Lower-extremity reflexes should also be tested in patients suspected of having cord injury or compression, including the patellar reflex for L4 and Achilles tendon for S1. As mentioned above, patients with cord compression from myelopathy or other sources will typically have hyperreflexia because of the diminished inhibition from upper motor neurons, although severe root compression may result in a decreased or even absent reflex arc. Reflexes should always be compared bilaterally and can be reported as

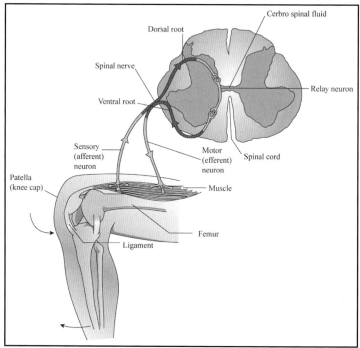

Figure 1-3. The stretch reflex arc begins with a muscle spindle responding to stretch and transmitting a signal via an afferent peripheral nerve to the spinal cord synapse. The signal is then relayed to the efferent nerve and on to the muscle fibers, eliciting a response. Upper motor neurons descend from the brain to modulate the reflexes.

decreased, normal, or increased. Numerical grading of reflexes is also common, where grade 0 is absence of reflex, grade 1 is hyporeflexia, grade 2 is normal, and grade 3 is hyperreflexia. Pathologic reflexes should also be evaluated to rule out upper motor neuron pathology (Table 1-3).

Gait

The influence of the cervical spine on gait is primarily related to the effects of myelopathy. Compression of the corticospinal tracts and posterior columns can result in notable balance and coordination impairments. Myelopathic patients will often have a wide-based, unsteady gait, and they should also be evaluated for the ability to toe-walk, heel-walk, and perform a toe-to-heel tightrope gait.

Table 1-3

PATHOLOGIC REFLEXES

Sign	Explanation
Hoffmann's sign	Snapping the middle finger distal phalanx downward results in spontaneous flexion of the ipsilateral thumb in a positive test
Babinski's sign	Scrape a pointed object along the lateral plantar border of foot; extension of the great toe and splaying is an abnormal reflex, while down-going toes are normal
Oppenheim test	Scrape a pointed object over the tibial crest; extension of the great toe and splaying is an abnormal reflex, as in Babinski's sign
Clonus	Repetitive flexion-extension wavering caused by rapid dorsiflexion of the ankle
Scapulohumeral reflex	Scapula elevates or humerus abducts in response to tapping on the scapular spine
Crossed adductor response	Tapping on the medial femoral condyle causes contralateral lower-extremity adduction

SPECIFIC TESTS

A multitude of additional tests are helpful in examining the cervical spine. Many of these have been outlined with written and pictorial descriptions in Table 1-4. Testing peripheral nerves for entrapment syndromes can also help distinguish between cervical spine and peripheral etiologies of nerve injury. Tinel sign (tapping on the nerve to elicit pain or numbness and tingling) should be tested at the cubital tunnel or Guyon canal for the ulnar nerve (C8), the carpal tunnel for the median nerve (C6), and the dorsal forearm for radial nerve compression (C7).

Table 1-4

HELPFUL HINTS AND SPECIFIC TESTS
FOR THE CERVICAL SPINE EXAMINATION

Test	Photograph	Description
Lhermitte's sign	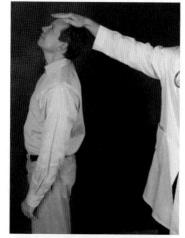	Shock-like sensation in trunk or extremities caused by axial load with flexion or extension of the neck; pain is believed to occur from reduction in foraminal or spinal canal space
Compression		Axial compression may reproduce symptoms related to nerve root compression, while gentle distraction may relieve those symptoms

(continued)

Table 1-4 (continued)

HELPFUL HINTS AND SPECIFIC TESTS
FOR THE CERVICAL SPINE EXAMINATION

Test	Photograph	Description
Hoffmann's sign		Snapping the distal phalanx of the middle finger downward causes involuntary flexion of the other fingers; this is a pathologic reflex caused by upper motor neuron dysfunction
Finger escape sign	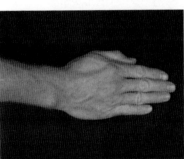 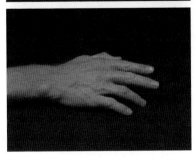	The patient holds the fingers adducted and extended; myelopathic patients will have their small finger and ring finger flex and abduct, usually in less than 1 minute

(continued)

Table 1-4 (continued)

HELPFUL HINTS AND SPECIFIC TESTS
FOR THE CERVICAL SPINE EXAMINATION

Test	Photograph	Description
Inverted radial reflex	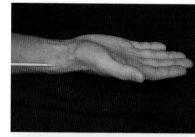	Hyperactive finger flexion occurs from tapping the distal brachioradialis tendon, indicating upper motor neuron dysfunction
Grip and release test	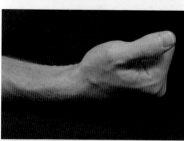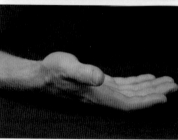	Normal patients should be able to make a fist and quickly release it 20 times in 10 seconds; myelopathic patients cannot do this that fast

(continued)

Table 1-4 (continued)

HELPFUL HINTS AND SPECIFIC TESTS FOR THE CERVICAL SPINE EXAMINATION

Test	Photograph	Description
Romberg sign		Positive if patient cannot maintain balance while standing with feet together, arms held forward, and eyes closed; indicates problem with dorsal column position sense
Spurling's test	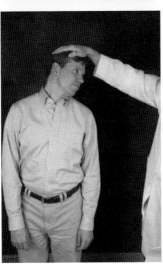	In this test, which can distinguish between shoulder pain and cervical radiculopathy, the neck is laterally flexed and rotated toward the side with pain; a positive test reproduces pain and indicates radiculopathy

Nerve compression alone can explain neurologic complaints, but patients may also experience a "double crush" phenomenon, in which peripheral nerve compression coexists with cervical radiculopathy, thus amplifying the nerve irritation and injury. Some patients may even undergo cervical spine surgery for radiculopathy that does not resolve after surgery and is later found to be due to peripheral nerve entrapments. This may be avoidable by conducting a thorough preoperative examination of the potential sites of peripheral nerve entrapment and referring the patient for electrodiagnostic studies if still unsure.

The cranial nerves should also be examined in anyone who has bulbar symptoms (ie, dysphagia, dysarthria, or facial weakness) or a history of head or spine trauma. Table 1-5 summarizes the function of cranial nerves.

In cases in which a patient's complaints do not seem to match the objective findings of the examination, a series of tests can help distinguish patients who may be exaggerating their pain symptoms. Waddell developed a validated series of tests that have become known as Waddell signs, including pain in nonanatomic distributions, pain out of proportion to stimulus, and exaggerated pain manifestations.[5] There are also 4 benign tests that are not sufficient to elicit a pain response and thus will be positive in patients who are exaggerating pain symptoms:

1. *Skin roll test*: With the patient standing or prone, gently roll the loose skin over the back and ask if radicular symptoms are elicited. No radicular symptoms should be caused by skin rolling.

2. *Twist test*: With the patient standing with hands on hips, gently rotate the torso. This will simulate spine motion, although the rotation actually occurs through the knees and should not cause back pain.

3. *Head compression*: Apply approximately 5 pounds of load to the top of the patient's head, which should not be enough to cause mechanical pain.

4. *Flip test*: A positive test occurs when a supine straight-leg raise causes pain but a seated straight-leg raise does not cause pain.

Table 1-5

CRANIAL NERVE FUNCTION

Cranial Nerve	Function
I Olfactory	Smell
II Optic	Vision
III Oculomotor	Eye movement
IV Trochlear	Superior oblique muscle
V Trigeminal	Mastication muscles and sensation for head/neck and external tympanic membrane
VI Abducens	Lateral rectus muscle
VII Facial	Facial muscles, sensation for ear and tympanic membrane, and taste anterior two-thirds of tongue
VIII Vestibulocochlear	Hearing and balance
IX Glossopharyngeal	Stylopharyngeus muscle, sensation for tongue and mouth, and taste posterior one-third of tongue
X Vagus	Motor and sensation in pharynx and larynx, and sensation from external ear
XI Spinal accessory	Motor to trapezius and sternocleidomastoid
XII Hypoglossal	Motor to tongue except palatoglossal

CONCLUSION

A detailed and complete examination of the cervical spine is the cornerstone of diagnosing spine pathology. Understanding the complexities of the motor, sensory, and reflex examinations and applying those in a thoughtful manner to the individual patient will result in consistent evaluations. Combined with the patient's history and essential imaging, a thorough physical examination will lead to appropriate diagnoses and treatment algorithms.

REFERENCES

1. Ono K, Ebara S, Fuji T, Yonenobu K, Fujiwara K, Yamashita K. Myelopathy hand: new clinical signs of cervical cord damage. *J Bone Joint Surg Br.* 1987;69(2):215-219.
2. Bland JH. Anatomy and pathology of the cervical spine. In: Giles LGF, Singer KP, eds. *Clinical Anatomy and Management of Cervical Spine Pain.* Philadelphia, PA: Elsevier; 2002.
3. White AA, Panjabi MM. Basic biomechanics of the spine. *Neurosurgery.* 1980;7(1):76-93.
4. Pullicino P. Bilateral distal upper limb amyotrophy and watershed infarcts from vertebral dissection. *Stroke.* 1994;25(9):1870-1872.
5. Waddell G, McCulloch JA, Kummell E, Venner RM. Nonorganic physical signs in low-back pain. *Spine.* 1980;5(2):117-125.

2

Physical Examination of the Thoracolumbar Spine

Brian T. Barlow, MD and Eric B. Harris, MD

INTRODUCTION

The thoracolumbar spine is a dynamic and heterogenous part of the spine. In the bipedal human, the thoracolumbar spine is the main weight-bearing structure of the spine. The thoracolumbar junction is a dynamic region responsible for much of our ability to flex, extend, and rotate our bodies. Anyone watching an Olympic gymnast perform can marvel at the amazing flexibility of the thoracolumbar spine. The large demands and stresses placed on the thoracolumbar

Rihn JA, Harris EB. *Musculoskeletal Examination of the Spine: Making the Complex Simple* (pp. 18-41).
© 2011 SLACK Incorporated.

spine as well as the surrounding muscles, tendons, and ligaments can predispose to injury. Evaluation of the injured back and thoracolumbar spine includes a thorough history, careful inspection, and provocative testing. Low back pain is exceedingly common; the lifetime prevalence of back pain in US adults is estimated to be as high as 70%.[1] In 2008, the US Bureau of Labor reported that 40.2% of all workplace injuries were back strains.[2] The general practitioner as well as the orthopedic surgeon will frequently see patients in the office who complain of low back pain. Proper treatment of the spine is an area fraught with misconceptions of patients who fail to improve. Successful treatment of the thoracolumbar spine hinges on accurate diagnosis, which starts with a comprehensive history and physical examination. However, before beginning a discussion of the physical examination, one must first define the terms with which the thoracolumbar spine is discussed.

DEFINITIONS

The following is a list of terms and definitions used to discuss pathology of the thoracolumbar spine:

Kyphosis–Outward curve of the thoracic spine, normal in the thoracic region. Too much kyphosis can also be termed as hunchback or Scheuermann disease.[3]

Lordosis–Inward curving of the lumbar spine, normal in the cervical and lumbar spine. Too much lordosis in the lumbar spine is termed swayback, and loss of the normal lumbar lordosis leads to flatback. Patients with chronic back pain often have loss of lumbar lordosis secondary to paraspinal muscle spasm.

Scoliosis–Lateral curving of the spine, which is always abnormal. Scoliosis is defined as a Cobb angle >10 degrees.[4] Calculating the Cobb's angle begins with identifying the superior and inferior transitional vertebra of the curve. The end or transitional vertebrae are the most superior and inferior vertebra that are least displaced and rotated, and have the maximally tilted end plate. Next, a line is drawn along the rostral (superior) end plate of the superior end vertebra, and a second line is drawn along the caudal (inferior) end plate of the inferior end vertebra. The angle between these 2 lines (or lines drawn perpendicular to them) is measured as the Cobb angle.[5] Although the Cobb angle is the most frequently used measurement in scoliosis, it reflects curvature only in the coronal plane and fails to account for sagittal imbalance or vertebral rotation. Therefore, the Cobb angle may not accurately demonstrate the 3-dimensional nature of the deformity.

Radiculopathy–Peripheral nerve lesion or disorder; mechanical compression of a nerve root as it exits the neural foramina of the spine. Radiculopathy is a preganglionic disorder of the spinal nerve. This condition often results in pain and can present with sensory and reflex disturbances. Radicular pain follows the anatomic course of the nerve and is often described as unilateral sharp, burning, or stabbing pain. The pain is often superficial and well-demarcated.[6]

Myelopathy–Central nervous system lesion or disorder. Myelopathy can be caused by a variety of etiologies including mechanical compression, infarction, malignancy, or infection. Symptoms often include bilateral weakness, decreased sensation, loss of proprioception, and hyperreflexia. Bilateral upper motor neuron symptoms are associated with myelopathy as opposed to unilateral lower motor neuron symptoms that are associated with radiculopathy. Myelopathy is frequently considered an orthopedic emergency requiring urgent surgical decompression.[7]

Lumbago–Low back pain or back strain; often due to muscular or ligamentous strain and can be secondary to trauma.

Spondylolisthesis–Horizontal translation of 2 vertebrae with respect to each other, frequently due to a lesion in the isthmus or pars interarticularis. One of the risks of spondylolysis is progression to spondylolisthesis.[6]

Spondylolysis–Defect or stress fracture in the pars interarticularis without slippage of one vertebral body onto another. The most common cause of low back pain in children and adolescents, spondylolysis is caused by repetitive hyperextension and occurs most commonly at L5.[6]

Spondylosis–Degenerative arthritis (osteoarthritis) of the spinal vertebra and related joints. If severe, it may cause pressure on nerve roots with subsequent pain or paresthesia in the limbs.[7]

Spinal stenosis–Narrowing of the spinal canal or neural foramina. Central stenosis results in compression of the thecal sac and consequently the spinal cord, and may present with myelopathy, conus medullaris syndrome, cauda equina syndrome, or radiculopathy.[7] Central stenosis has many etiologies such as central herniated disk, hematoma, infection, or spondylolisthesis. Lateral stenosis produces compression of the nerve roots and therefore radiculopathy. Lateral stenosis can be caused by overgrowth of the facet joint in the lateral recess or by a lateral herniated disk in the intervertebral foramen.[6]

History

As mentioned above, low back pain is an exceedingly common complaint. It is important to obtain an accurate history when examining a patient with a new complaint of back pain. The importance of characterizing an accurate timeline of the onset and development of pain cannot be overstated. The differential diagnosis for acute versus chronic pain are markedly different, and the chronology of pain changes the likelihood of a specific diagnosis. In addition, one must further clarify the history of back pain by eliciting the following information: provoking movements or positions, response of pain to anti-inflammatory medications, limitation of daily activities, quality (ie, sharp, dull, or electrical/shooting), severity, and associated symptoms (eg, paresthesias, gait disturbance, and loss of bowel or bladder function). Gathering an accurate history of the use of complementary and alternative medicines or therapies also has gained significance in the past few decades.

Symptoms of unremitting back pain, bowel or bladder incontinence, and new back pain in patients with a history of malignancy require immediate evaluation.[1] These findings are associated with progressive neurologic deficiency that may be amenable to surgical decompression.

Inspection

The following is a list of steps to follow when performing an inspection of the spine:

1. Observe the patient's back without any covering clothes (Figure 2-1). Maintain appropriate draping as necessary, but one of the most important tests for the thoracolumbar spine is observation of a **standing** patient. Look for

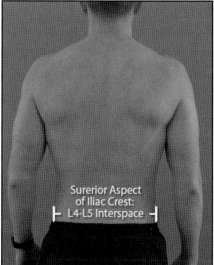

Figure 2-1. Observe the patient's back, looking for differences in muscle bulk, skin changes, and muscle atrophy. Palpate the superior aspect of the iliac crest, the horizontal connecting line corresponds to the L4-L5 interspace.

Surerior Aspect of Iliac Crest: L4-L5 Interspace

postural abnormalities in both the coronal and sagittal planes. Also, observe the patient's muscle bulk and tone, posture, and any signs of trauma.

2. The thoracic spine should have a natural kyphosis. The thoracic spine is naturally stiff from the vertebrocostal joint connecting the spine to the ribs, the extensive musculature attachments to the thoracic vertebrae, and due to the orientation of the zyphoapophyseal (facet) joints.[5] The facets in the thoracic spine are principally coronal, which lends to good lateral bending but less flexion and extension. A sharp kyphosis of the thoracic spine seen in older adults is termed a Gibbous deformity. Observe the thoracic spine for abnormal superficial lesions; for example, a tuft of midline hair in the lower back (known as "faun's beard") may indicate an undiagnosed spina bifida.[8]

3. The lumbar spine has a natural lordosis. The lumbar spine is particularly dynamic in flexion and extension due to the sagitally oriented facet joints.[9] Patients should be observed both at the extremes of flexion (have the patient touch their toes) as well as extension. Normal extension should be between 5 and 15 degrees, whereas normal lumbar flexion is 60 to 90 degrees (Figures 2-2, 2-3, 2-4, and 2-5; Table 2-1).[3]

Figure 2-2. Normal lumbar extension is 5 to 15 degrees.

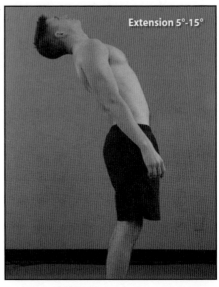

Figure 2-3. Normal lumbar flexion is 60 to 90 degrees.

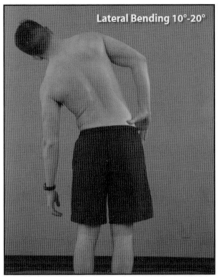

Figure 2-4. Normal lumbar lateral bending is 10 to 20 degrees.

Figure 2-5. Normal lumbar rotation is 5 to 15 degrees.

Table 2-1

LUMBAR AND THORACIC SPINE RANGE OF MOTION

Motion	Range
Lumbar spine	
Flexion	60 to 90 degrees
Extension	5 to 15 degrees
Lateral bending	10 to 20 degrees
Rotation	5 to 15 degrees
Thoracic spine	
Forward flexion	50 degrees
Extension	20 degrees
Lateral flexion	30 degrees
Rotation (total arc)	75 degrees

Adapted from American Medical Association. *Guides to Evaluation of Permanent Impairment.* 6th ed. Chicago, IL: American Medical Association; 2007.

4. The provider should pay particular attention to the muscle bulk of the legs when evaluating the lumbar spine (Figure 2-6). Unilateral atrophy of a quadriceps muscle may be associated with an L3-L4 radiculopathy, whereas unilateral atrophy of the calf suggests an L5-S1 radiculopathy.

5. Proper evaluation of the thoracolumbar spine requires observation of the patient's gait. A wide-based, waddling gait is associated with myelopathy. An antalgic or painful gait may be associated with radiculopathy.[6] In the radiculopathy work-up, the provider should observe the patient's ability to toe- and heel-walk. Inability to toe-walk or rapid fatigability with toe raises suggests S1 radiculopathy. Inability to heel-walk suggests L5 radiculopathy. Leg pain while standing or walking that

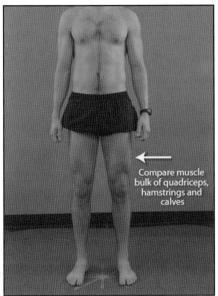

Figure 2-6. Observe the standing patient from all sides, taking care to identify any muscle bulk asymmetry or atrophy, especially in the lower extremities.

Compare muscle bulk of quadriceps, hamstrings and calves

is relieved by leaning forward (eg, leaning on a shopping cart with improvement in back and leg pain) is suggestive of spinal stenosis.

PALPATION

Spinous processes are oriented horizontally in the lumbar spine and slope downward in the thoracic spine. The most prominent thoracic spinous process is at T1.[10] The spinous processes of the lumbar spine are palpable, but it is difficult to determine the level though palpation alone. The supraspinous ligament is a palpable thick strip of connective tissue that attaches between each spinous process.[3] The examiner should palpate the spinous processes of the entire spine with particular attention to step-offs and point tenderness, which suggests fracture. Palpate the sacral spinous processes; the coccyx is the most distal aspect of the spine and is palpable at the superior aspect of the gluteal cleft.

Inspection of the lumbar spine also requires inspection of the pelvis. A useful surgical landmark is the horizontal line

connecting the superior aspect of the iliac crests corresponding to the L4-L5 interspace.[10] Palpate the greater trochanter
and the ischial tuberosities; the sciatic nerve runs between
these 2 structures. The horizontal line connecting the posterosuperior iliac spine correlates to the S2 level.[10]

Paraspinal muscles flank the posterior spine bilaterally
and are innervated by the posterior rami of lumbar nerves.
The paraspinal muscles are generally split into 2 levels: the
superficial layer consisting of the sacrorspinalis and erector
spinae and the deep layer consisting of the multifidus and
rotator muscles. Anatomy texts show many named muscles,
but ultimately these muscles are not dissected individually
during the posterior approach. The paraspinal muscles are
separated from the spinous process and retracted laterally
from the spine during the posterior midline. The innervation
of the paraspinal muscles is diffuse from each vertebral level,
therefore dissection off the spinous process, lamina, and facet
joints does not result in deinnervation.[10]

STRENGTH TESTING

Muscle examination of the thoracolumbar spine is accomplished through lower-extremity muscle testing. The examiner must bear in mind that there is naturally some overlap
between different levels. The L1-L2 nerve roots innervate the
abductors of the hip and are tested via leg abduction. The L2-
L3 nerve roots innervate the iliopsoas muscles and are tested
via hip flexion. The L4 nerve root innervates the quadriceps
and tibialis anterior; L4 is tested via knee extension and foot
dorsiflexion. The L5 nerve root innervates the extensor hal-
licus longus and is tested by great toe dorsiflexion. The S1
nerve root innervates the gastrocnemius-soleus complex and
is tested via foot plantarflexion.

SENSORY TESTING

When assessing thoracolumbar sensation, a rule of thumb
is to test at least one modality from both the spinothalamic
tracts and dorsal columns. Spinothalamic tracts provide pain

and temperature information. The second-order neurons are found in the dorsal horn of the spinal cord shortly after entering the spinal cord, often within 1 to 2 levels. The second-order neurons then cross the midline in the ventral white commissure of the spinal cord and ascend the spinal cord contralaterally.[11] The dorsal columns (cuneate fasciculus and gracile fasciculus) provide light touch, proprioception, and vibration sensation. These fibers ascend the spinal cord ipsilaterally, and the second-order neurons are located in the gracile and cuneate nuclei of the caudal medulla. The secondary neurons then cross the medulla as the medial lemniscus.[11]

Sensory testing of the thoracic and lumbar spine is quite different. The thoracic spine sensory dermatomes are a series of bands on the trunk. Some of the important landmarks include the manubrium, which corresponds to T4; the xiphoid process is innervated by T7; the umbilicus is reliably innervated by T10; and finally, the inguinal crease sensation is provided by T12.[3] Other than for specific thoracic complaints, many clinicians do not routinely test the thoracic sensory dermatomes.

Unlike the thoracic dermatomes, evaluation of the lumbar dermatomes is routinely considered an essential part of any visit for back or lower-extremity complaints. The L1-L3 nerve roots innervate the anterior thigh and the superior aspect of the knee. L4 innervates the inferior knee and the medial lower leg and medial edge of the foot. The sharp anterior aspect of the tibia is the dividing line between the medial and lateral leg. L5 innervates the lateral side of the lower leg and the dorsum of the foot. S1 innervates the posterior thigh, calf, lateral edge, and plantar surface of the foot. S2-S4 innervates concentric circles of the perineum.[3]

REFLEXES

The simple reflex arc is a sensorimotor loop that can function without using the ascending/descending white matter tracts, although these higher cortical systems modulate the reflex. The simple myotactic reflex is composed of an afferent limb, which includes a muscle spindle receptor and the nerve traveling to the dorsal root ganglion, and an efferent limb, which includes the ventral horn motor neuron that innervates the striated muscle of interest.[12] The muscle spindle receptor is triggered by sudden tendon stretch (when the hammer hits the tendon) and relays this signal via the afferent fibers to the dorsal root ganglion. The reflex arc is then conveyed to the ventral horn motor neuron via the anterior horn cell of the spinal cord. Finally, the ventral horn motor neuron fires, causing the muscle to contract.[11]

It is important to recognize that the cerebral cortex provides modulates to the deep reflexes. Upper motor neuron lesions are associated with exaggerated or spastic reflexes due to loss of cortical inhibition of the reflex arc.[11] In addition, loss of upper motor neuron inhibition can lead to reflex phenomena such as cross-over and failure to extinguish. Cross-over is described as reflex stimulation while testing the contralateral side; for example, striking the patella tendon of one side causes the contralateral quadriceps to contract. Cross-over is associated with an upper motor neuron lesion. A reflex that fails to extinguish is also associated with an upper motor neuron lesion. Normally, repetitive striking of a tendon such as the patellar tendon produces a diminishing response after several hammer strikes. In a patient with an upper motor neuron lesion, the reflex fails to extinguish no matter how many times the reflex is repeated. The major reflexes of the lumbar spine[6] include the L4 reflex (patellar): tap the patellar ligament, which causes knee extension via contraction of the quadriceps muscles; and the S1 reflex (Achilles): tap the Achilles tendon, which causes plantarflexion of the foot via contraction of the gastrocnemius soleus.

The superficial reflexes of the thoracolumbar spine include the abdominal, cremasteric, and anal wink reflexes. These reflexes are mediated through cutaneous sensation, which is transmitted to the cerebral cortex.

The cremasteric reflex is mediated by L1-L2 and refers to elevation of the ipsilateral scrotum with light stroke of the inner thigh. The abdominal reflex is the only reflex of the thoracic spine and correlates to T7-T10 in the upper abdomen and T10-L1 in the lower abdomen. Specifically, a light scratch in each quadrant of the abdomen triggers movement of the umbilicus towards that quadrant. This reflex is subtle and often difficult to test in many patients. The anal wink reflex refers to anal contraction with light touch to the perianal skin. The S1-S3 nerve roots mediate this reflex.

The bulbocavernosus reflex is of particular importance in acute spinal cord injury. Specifically, the bulbocavernosus reflex is contraction of the anal sphincter in response to stimulation of the bladder trigone through either squeezing the glans of the penis, tapping the mons pubis, or gently tugging on a Foley catheter. The absence of this reflex signifies spinal shock, which usually lasts approximately 24 hours after an acute spinal cord injury.[5] Spinal shock is defined as cord dysfunction that is due to physiologic compromise and the spine "shutting down" in response to injury, and may be reversible. This is contrasted by structural compromise, which is not reversible. Some of the deficits of the cord injury may return after the spinal shock resolves, but it is difficult to say which modalities. Return of the bulbocavernosus reflex signifies the end of spinal shock, and any lesions present at this time are essentially permanent.[5]

SPECIFIC TESTS

This section is dedicated to the provocative tests of the thoracolumbar spine. These tests are important adjuvants to the aspects of the examination described above. The value of an accurate history; careful inspection; and accurate motor, sensory, and reflex testing are of the utmost importance when examining the thoracolumbar spine. The value of these tests is variable; some are key in the examination of thoracolumbar spine, whereas others are more of historical or academic interest.

Babinski's Test

Babinski's test (Figure 2-7) is a classical test for an upper motor neuron lesion and is an essential part of the basic neurologic examination. The test is conducted by running a sharp instrument along the lateral plantar aspect of the foot. A negative response is plantarflexion of the toes. A positive test is dorsiflexion of the great toe and splaying of the other toes. A positive test signifies an upper motor neuron lesion in adults for a number of reasons including spinal trauma, brain damage, or brain and spinal malignancy.[13] One should note that a positive test is a normal finding in children <1 year because their spinal cord has not completely myelinated.

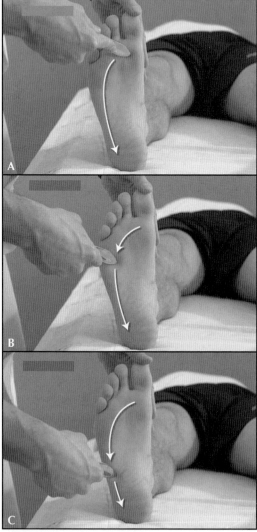

Figure 2-7. (A–C) Babinski's test is conducted by running a sharp instrument along the lateral plantar aspect of the foot. A negative response is plantarflexion of the toes, and a positive test is dorsiflexion of the great toe and splaying of the other toes.

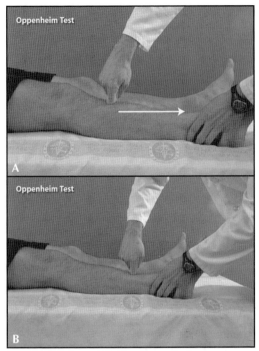

Figure 2-8. (A and B) The Oppenheim test is performed by running a knuckle along the anterior tibial spine. A positive response is dorsiflexion of the great toe and splaying of the other toes.

Oppenheim Test

The Oppenheim test (Figure 2-8) is an alternative to Babinski's test, and a positive Oppenheim test is equal in significance to a positive Babinski's test. The Oppenheim test is performed by running a knuckle along the anterior tibial spine.[13] A positive Oppenheim test is identical to a positive Babinski's test and is dorsiflexion of the great toe and splaying of the other toes.

Brudzinski and Kernig Tests

Brudzinski and Kernig tests are used to determine the presence of increased intrathecal pressure frequently ascribed to meningeal infection. The Brudzinski test is performed by having the patient flex both knees up to his or her chest. A positive finding is increased pain in the back and neck, and indicates increased intrathecal pressure.[3] The Kernig test is performed by having the patient use both hands to flex the neck and bringing his or her chin to the chest. Pain in the cervical spine, lower back, or radiating to the legs is indicative of meningeal irritation.[3] (These 2 tests are mandatory when assessing a patient for meningitis.)

Straight-Leg Raise Test for Disk Herniation

Lumbar disk pathology is a common reason for operative management of back and leg pain. Spinal nerves in the lumbar spine are either exiting or traversing. In the thoracolumbar spine, the exiting nerve root turns and exits just below the pedicle of the caudal vertebra. For example, the L4 nerve root exits at the L4-L5 interspace. The L5 nerve root is the traversing root at that level (L4-L5). A herniated vertebral disk is the consequence of a tear in the anulus fibrosus, leading to extravasation of the nucleus pulposus.[9] The extruded nucleus pulposus compresses the nerve root. The perforation most commonly occurs just lateral to the posterior midline and impinges on the traversing nerve root. This spot just off midline is where the posterior longitudinal ligament is the weakest. This type of herniation accounts for 95% of all disk herniations and is most common at the L4-L5 and L5-S1 levels.[7]

A variant of disk herniation is the far-lateral disk where the herniated disk impinges on the exiting nerve root. In this instance, the herniation causes symptoms often ascribed to the level below where the actual lesion is. For example, a far-lateral disk herniation at L5-S1 causes L5 radiculopathy, which would commonly be ascribed to a conventional L5-S1 disk herniation.[6] The most common complaint with disk herniation is radicular pain shooting down a unilateral leg. Table 2-2 lists common lumbar disk syndromes.

The most reliable and ubiquitous test used in diagnosing disk herniation symptoms is the straight-leg raise test (Figure 2-9). To perform the straight-leg raise test, the examiner raises the leg of interest until it becomes painful. If the patient has pain from 35 to 70 degrees, this is considered a positive test.[7]

Table 2-2

LUMBAR DISK SYNDROMES AND CLINICAL FINDINGS

Level	Affected Root	Sensory Loss	Motor Loss	Reflex Loss
L1-L3	L2, L3	Anterior thigh	Hip abductors and hip flexors	None
L3-L4	L4	Medial calf and medial foot	Quadriceps and tibialis anterior	Patellar
L4-L5	L5	Lateral calf and dorsum of foot	Extensor hallicus longus	None
L5-S1	S1	Posterior calf and lateral and plantar foot	Gastrocnemius and soleus	Achilles
S2-S4	S2-S4	Perianal	Bowel or bladder continence	Cremasteric

Adapted from Miller MD. *Review of Orthopaedics.* 4th ed. Philadelphia, PA: WB Saunders Co; 2004.

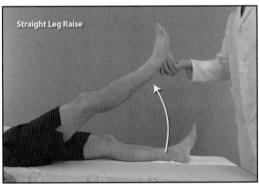

Figure 2-9. The straight-leg raise test is performed by slowly raising the patient's on the affected side until pain is elicited in the back and is radiating down the examined leg. Pain often occurs between 35 and 70 degrees from horizontal.

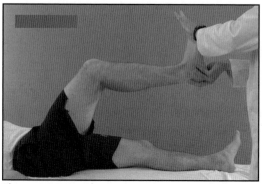

Figure 2-10. For the Lasègue test, a straight-leg test is performed. After pain is elicited, the examiner drops the leg several degrees until pain is relieved and then dorsiflexes the foot. Reproduction of pain is a positive finding.

Lasègue and Well-Leg Raise Tests

A variation of the straight-leg raise test is the Lasègue test (Figure 2-10) in which the examiner drops the leg several degrees after pain is elicited and then dorsiflexes the foot. Reproduction of the pain with dorsiflexion of the foot confirms the presence of an ipsilateral radiculopathy.[13]

A complementary test is the well-leg raise test. In this test, the examiner raises the unaffected straight leg, and pain is elicited in the contralateral leg. A positive well-leg raise test is highly suggestive of nerve root entrapment.

Sciatic pathology is a common complaint of patients with back pain, and there are many tests of sciatic irritation. Sciatica is defined as leg pain secondary to mechanical compression or irritation of the sciatic nerve. Sciatica is considered to be irritation of the nerve distal to the nerve root.[7] There is a fine distinction between radicular pain and sciatic pain, which often have overlapping symptoms. Some of the commonly used tests for sciatica are described on the following pages.

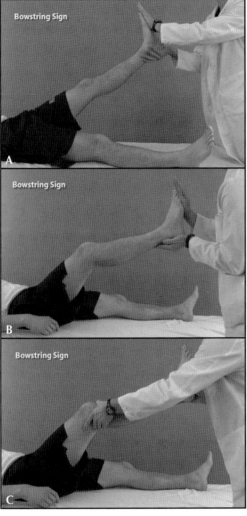

Figure 2-11. (A) For the Bowstring sign, a straight-leg raise is performed. (B) Pain is relieved with knee flexion by taking tension off the sciatic nerve. (C) Placing compression on the popliteal fossa causing the return of symptoms is a positive finding.

The Bowstring sign (Figure 2-11) is a variation of the straight-leg raise. The examiner performs a straight-leg raise and the pain is relieved with knee flexion by taking tension off the sciatic nerve. Placing compression on the popliteal fossa causing the return of symptoms is a positive finding.[14]

Figure 2-12. The axial load test is performed by applying mild load (<5 pounds) to the top a the patient's head. Provocation of back pain is a sign of nonorganic back pain.

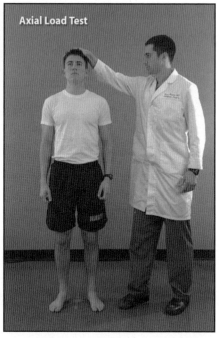

Axial Load Test

Occasionally, the orthopedic surgeon will have patients he or she suspects may be exaggerating their symptoms for secondary gain. This situation is prevalent enough that validated examinations are used for confirming the malingering. A series of tests called the Waddell signs include 23 signs and symptoms of nonorganic pain.[13] The following is a list of some of the more important ones[1]:

- Pain with palpation of the tip of the tailbone
- Tenderness with light touch in nonanatomic distributions
- Incongruent distribution of pain
- Reproduction of pain with lightweight (<5 pounds of pressure) axial loading (Figure 2-12)
- Pain with rotation of the pelvis with hands on the hips (rotation actually occurs in the knee and ankle joints)

Flip test (Figure 2-13) where the pain elicited during a supine straight-leg raise is not reproduced with the patient in a sitting position.

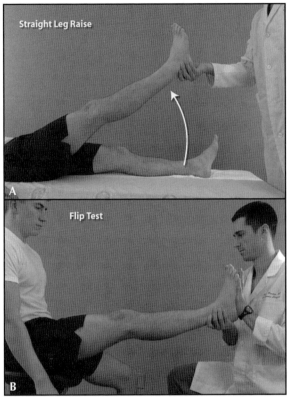

Figure 2-13. (A and B) In the Flip test, a straight-leg raise is performed first with the patient supine and then again with the patient sitting. If pain is not reproduced, then the results are questionable.

Another important test for malingering is the Hoover test (Figure 2-14). In this examination, the patient places both heels in the examiner's hands and is asked to lift one leg. If the patient cannot lift the "affected" leg, the examiner should feel downward pressure in the opposite leg. If there is no downward pressure in the leg opposite the one being examined, then the patient may not be actually attempting to cooperate with the examination.[13]

Although a thorough examination should be performed in all patients complaining of pain, these tests may suggest a cause that is not due to anatomic abnormality but rather to secondary gain, psychiatric condition, or Munchausen syndrome.

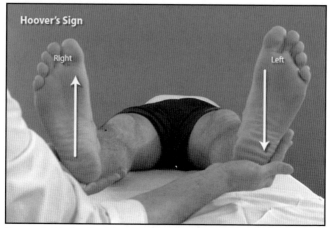

Figure 2-14. In the Hoover test, the patient is asked to lift the affected foot from the examiner's hand. The examiner should feel downward pressure in the unaffected contralateral foot if the patient is cooperating with the examination.

CAUDA EQUINA AND CONUS MEDULLARIS SYNDROMES

Finally, a discussion of the thoracolumbar spine would not be complete without defining and distinguishing the syndromes of the distal spinal cord, the conus medullaris, and the terminal nerve roots, the cauda equina. Cauda equina syndrome is a serious condition caused by compression of the lumbar and sacral nerve roots after the conus medullaris, which is the terminal end of the spinal cord. Symptoms include unilateral radicular pain (sharp, stabbing pain) of the posterior thigh and leg; unilateral lower extremity numbness of the buttock, posterior thigh, and plantar aspect of the foot; unilateral saddle anesthesia; occasional urinary retention; and infrequent loss of bowel and bladder continence.[7] Muscle atrophy of the ipsilateral calf and loss of patellar and Achilles reflex on the affected side are other important characteristics of cauda equina syndrome. Cauda equina syndrome is usually a spine emergency requiring immediate radiographic evaluation. The etiology is variable and can be caused by a nerve root hematoma, malignancy, ependymoma, or massive disk herniation.

Conus medullaris syndrome, on the other hand, is compression of the terminal end of the cord affecting the sacral cord elements. Distinguishing characteristics from cauda equina syndrome include pain that is usually bilateral and less severe. Normal or slightly decreased reflexes help distinguish conus from cauda. Hallmark symptoms also include bilateral saddle anesthesia, severe dysfunction of bowel and bladder function, and prominent sexual dysfunction.[13] The muscle changes are less pronounced compared to cauda equina syndrome. The etiology is similar to cauda equina but occurs higher in the spinal cord. Similar to cauda equina syndrome, conus medullaris is a neurologic emergency that requires immediate investigation.

CONCLUSION

The thoracolumbar spine is an area of specialized function and great flexibility. The thoracolumbar spine is subjected to significant stress, and many patients suffer from back pain. The history and physical exam are paramount in the evaluation of back pain and the starting point from which appropriate further diagnostic work-up is begun. Most patients with back pain do not require surgical intervention for improvement in their symptoms; but the history and physical examination are key in determining which patients may benefit from spine surgery.

After a comprehensive history and examination are obtained, the proper diagnostic imaging can be ordered including magenetic resonance imaging and computed tomography. Given the high cost, radiation exposure, and limited resources in today's hospitals, the burden of responsibility falls onto the orthopedic provider to determine which patients require the additional imaging. Imaging studies must be a confirmatory adjunct to the history and physical examination, not the other way around. Spine surgeons must operate on the patient, not the imaging. The thoracolumbar spine is a complex structure of bony, neurologic, and soft-tissue elements, all of which can cause a patient symptoms. The proper diagnosis is dependent on thorough history and accurate physical examination.

REFERENCES

1. Shen FH, Samartzis D, Andersson GB. Nonsurgical management of acute and chronic low back pain. *J Am Acad Orthop Surg.* 2006;14(8):477-487.
2. United States Department of Labor. Nonfatal occupational injuries and illnesses requiring days away from work, 2008. http://www.bls.gov/iif/oshcdnew.htm. Accessed November 24, 2009.
3. Thompson JC. *Netter's Concise Atlas of Orthopaedic Anatomy.* Philadelphia, PA: Saunders; 2002.
4. McCarthy JJ, D'Andrea LP, Betz RR, Clements DH. Scoliosis in the child with cerebral palsy. *J Am Acad Orthop Surg.* 2006;14(6):367-375.
5. Vaccaro AR, Kim Dh, Brodke DS, et al. Diagnosis and management of thoracolumbar spine fractures. *Instr Course Lect.* 2004;53:359-373.
6. Miller MD. *Review of Orthopaedics.* 4th ed. Philadelphia, PA: Saunders; 2004.
7. Wiesel SW, Delahay JN. *Essentials of Orthopaedic Surgery.* 3rd ed. New York, NY: Springer; 2007.
8. Hedequist D, Emans J. Congenital scoliosis. *J Am Acad Orthop Surg.* 2004;12(4):266-275.
9. Eastlack RK, Bono CM. Fractures and dislocations of the thoracolumbar spine. In: Bucholz RW, Heckman JD, Court-Brown C, eds. *Rockwood and Green's Fractures in Adults.* 6th ed. Philadelphia, PA: Lippincott Wiliams & Wilkins; 2009:1377-1411.
10. Hoppenfield S, deBoer P. *Surgical Exposures in Orthopaedics: The Anatomic Approach.* 3rd ed. Philadelphia, PA: Lippincott Williams & Wilkins; 2003:247-344.
11. Fix JD. *High-Yield Neuroanatomy.* 3rd ed. Philadelphia, PA: Lippincott Williams & Wilkins; 2005.
12. Park AE, Boden SD. Form and function of the intervertebral disc. In: Einhorn TA, O'Keefe RJ, Buckwalter JA, eds. *Orthopaedic Basic Science: Foundations of Clinical Practice.* 3rd ed. Rosemont, IL: American Academy of Orthopaedic Surgeons; 2007:259-264.
13. Child Z. *Basic Orthopedic Exams.* Philadelphia, PA: Lippincott, Williams & Wilkins; 2007.
14. Patel M, Shah K. *Back: Cervical and Thoracolumbar Spine. Rakel Textbook of Family Medicine.* 7th ed. Philadelphia, PA: Saunders; 2007.

II

General Imaging

3

Degenerative Conditions of the Spine

Christopher Loo, MD, PhD; Harvey E. Smith, MD;
and Bradley K. Weiner, MD

Introduction

Degenerative spinal disorders encompass a wide variety of clinical entities that vary based on their epidemiology, pathophysiology, spectrum of clinical symptoms, work-up, and treatment. Degenerative conditions are generally progressive in nature and slow in their onset. Typically, these conditions may have an acute exacerbation that precedes the initial presentation to the orthopedist. In addition, acute symptoms can arise in the setting of chronic degenerative conditions.

Rihn JA, Harris EB. *Musculoskeletal Examination of the Spine: Making the Complex Simple* (pp. 43-49).
© 2011 SLACK Incorporated.

Diagnostic imaging of degenerative conditions necessitates clinical judgment based on the interpretation of symptoms and physical findings. In contrast to the setting of acute spine trauma, imaging for degenerative conditions generally occurs in the office or clinic setting. No given imaging technique is the sole basis for diagnosis, but rather imaging findings must be placed in the proper clinical context to arrive at an accurate diagnosis and to formulate an appropriate treatment plan.[1]

The initial work-up of a patient with a degenerative spine condition includes a thorough clinical history and physical examination; pain is usually the presenting symptom. Additional symptoms include a limitation in physical activities, feelings of numbness or tingling, and changes in gait. Some common entities to consider in the differential diagnosis of degenerative conditions of the spine include degenerative disk disease, facet joint osteoarthritis, herniated disk, spondylolysis, spondylolisthesis, and spinal stenosis. Careful attention should be given to the consideration of systemic conditions such as ankylosing spondylitis, rheumatoid arthritis, and malignancy in the differential diagnosis. If a systemic condition is suspected, laboratory work-up including erythrocyte sedimentation rate, C-reactive protein, and complete blood cell count with differential may be necessary.

Imaging of degenerative spine conditions encompasses plain radiography, computed tomography (CT), and magnetic resonance imaging (MRI). Additional imaging such as a bone scan may be required in the case of either a primary or metastatic malignancy.[1]

RADIOGRAPHY

Plain radiographs of the spine are usually the first choice because they are low cost and readily available. The advantages of radiography include ease of acquisition, cost, and time. Plain radiographs including anterorposterior and lateral views of the area of spine of interest are usually ordered first. Other imaging views include flexion-extension and oblique views. Images should be obtained with the patient standing, as under loading a spondylolisthesis may be more apparent than on a supine image.

Plain radiography can be used to assess bony alignment, structure, disk and vertebral body height, gross bone density and architecture, and the presence of osteophytes. Plain radiography is excellent for detecting bony changes, which are affected in conditions such as ankylosing spondylitis, rheumatoid arthritis, compression fractures, cancer, spondylosis, and spondylolisthesis. Certain conditions such as ankylosing spondylitis require visualization of specific joints such as the sacroiliac joints, which can be better visualized on angled views.[2,3] Careful attention should be exercised when evaluating plain radiographs since radiographic findings may be incidental and not an actual contributor to the patient's clinical condition. In addition, certain diseases such as malignancy may not be evident on routine radiographs until advanced stages.

Clinical studies have consistently failed to demonstrate a significant relationship between spinal degeneration and back pain based on data from plain radiographic testing. Poor imaging quality has been cited as a potential reason.[4] Therefore, in most cases plain radiography is the first step, followed by further advanced imaging if clinically indicated.[1] Correlation may exist between degenerative findings on plain radiographs and symptoms, but the clinician must use clinical judgment in inferring causality. The disadvantages of plain radiographs include poor visualization of soft-tissue structures and low sensitivity and specificity compared to CT and MRI.

COMPUTED TOMOGRAPHY

CT is a technique that uses X-rays to generate multiplanar images. In contrast to radiography, CT optimizes delineation of bony architectural details that are particularly relevant to degenerative disease. CT has high sensitivity and specificity in visualizing the foraminal nerves and the surrounding bony pathology, due to the surrounding fat providing natural contrast. However, CT is less effective in visualizing intra- and extrathecal nerve root compression. CT may also have a role (depending on the clinical scenario) in the imaging of compression fractures, disk herniations, degenerative disk disease, ankylosing spondylitis, rheumatoid arthritis, nerve root compression, stenosis, and malignancy.[1]

CT findings in the setting of degenerative disk disease may include intervertebral disk space narrowing, facet joint changes, spondylolisthesis, spinal stenosis, end-plate irregularity and sclerosis, vacuum phenomenon in the disk space (ie, air within the disk space that is evident on CT as a dark black stripe within the degenerated disk), and subchondral cysts, as well as calcification of the joint capsule, vertebral end plates, and ligaments.[5-7] Evaluation of the density of the paraspinal muscles can also give an indication of the extent of degenerative disease, as extensive degeneration may be associated with muscle atrophy.

MAGNETIC RESONANCE IMAGING

MRI has achieved great sensitivity in depicting degenerative changes. Traditional images include T1- and T2-weighted images in the axial and sagittal planes. In a T1-weighted image, the cancellous bone is brighter (ie, white) compared to the surrounding fluid and soft-tissue structures. In contrast, the fluid and disk are brighter in a T2-weighted image due to the enhancement of tissue structures containing more fluid.[8] MRI has good sensitivity for the detection of nonmechanical causes of low-back pain including tumor and infection.[1] However, in cases of mechanical back pain, MRI is sensitive for detecting degenerative changes but does not accurately predict the etiologies.[8]

The advantages of MRI include its noninvasiveness (relative to CT myelography), multiplanar imaging capacity, capability to detect intrinsic abnormalities of the cervical cord, and visualization of segments distal to a total myelographic block. MRI can provide excellent soft-tissue contrast images of the intervertebral disk, ligament, nerve, and cerebrospinal fluid.

Evidence of degenerative disk disease on MRI includes reduced signal intensity of the disk ("black disk"), reduced disk space height, the presence of annular tears, and the presence of a disk herniation. Tears within the disk may be enhanced with contrast. MRI can detect disk herniations with an extremely high diagnostic accuracy (ie, up to 100%),[1] and the effect of the herniated disk on nerve root compression can be assessed as well.

Intervertebral disks can be classified on the basis of MRI into 5 stages of degeneration. Classification systems to describe intervertebral disk degeneration include the Pfirrmann[9] and Thompson[10] scales. The Thompson classification system of intervertebral disk degeneration is used in clinical and experimental investigations. It is based on the number and type of fissures seen in the disk, the distinctness of the boundary between the collagenous and the cartilaginous portions of the disk, and the height of the disk on MRI. The scale was originally designed for use with anatomic slices. The Pfirrmann scale has a similar classification but was designed specifically for T2-weighted MRI. The Thompson and Pfirrmann scales have disadvantages including subjective differentiation between stages, ambiguity regarding degeneration versus age-related changes, and discontinuous scales. Type I, II, and III changes have been described[11,12] in end-plate changes by the American Society of Neuroradiology.[13] However, these morphologic changes in the disk and in the end plate do not equate to specific clinical symptoms.

A study conducted by Thornbury et al[14] showed that MRI has approximately the same sensitivity and specificity as CT (88% to 94% and 57% to 64%, respectively). In addition to imaging degenerative and herniated disks, MRI has also been used to image central stenosis, annular tears, and nerve root impingements. Other more advanced methods of MRI include functional, dynamic, T2-relaxometry, and magnetic resonance spectroscopy.[8]

COMPUTED TOMOGRAPHY MYELOGRAPHY

Although computed tomography myelography (CTM) was used in earlier years along with MRI to evaluate degenerative spinal disorders, its use has diminished in comparison to the use of MRI due to recent advances. Currently, MRI is considered to be ideal for patients with cervical myelopathy and radiculopathy. CTM is usually reserved for patients in whom MRI results were ambiguous or technically suboptimal or in clinical scenarios in which MRI is contraindicated (eg, patient with a pacemaker) or limited due to artefact (such as in the setting of prior instrumentation).

CTM is still used by some surgeons for further information about cord compression or foraminal stenosis as a result of bony lesions. It is also used to assess for the degree or nature of compression; thus, when evaluating spondylotic myelopathy, some consider CTM to provide more information than MRI. In addition, MRI has poor interobserver correlation in detecting bony lesions.[15]

Due to the size of the neural foramen and relative lack of epidural fat (which provides natural contrast) in the cervical spine, it is not easy to obtain reliable results regarding foraminal compression in the cervical spine. Nguyen-minh et al[16] reported that it is difficult to use MRI for characterizing a lateral lesion as a bone or soft tissue. Karnaze et al[17] reported that CTM was more accurate than MRI for spondylosis. Modic et al[11] noted that CTM was more accurate than MRI in radiculopathic cases. However, both Karnaze and Modic's work was before the widespread implementation of higher-field MRI systems and the routine use of volumetric coils, advances that have significantly improved MRI resolution and signal-to-noise ratio.

Song et al[15] also showed that CTM yielded more reliable results, but in nerve root compression, MRI was more reliable. Although CTM is useful for preoperative planning, it is limited as a primary diagnostic tool due to different interpretations, especially regarding nerve root compression. Shafaie et al[18] found that the degree of concordance between CTM and MRI was poor in lateral recess lesions. Discrepancies were noted in the differentiation of soft-tissue compression versus spondylotic osteophytes.

CONCLUSION

Degenerative spine disease encompasses a variety of clinical entities. The use of imaging modalities to depict changes to arrive at an accurate diagnosis is important for devising a proper treatment plan. Traditionally, the "go-to" method has been plain radiography due to its ubiquitous nature and low cost. However, the limitations of plain radiography in combination with more recent advances in imaging including CT, MRI, and CTM, have allowed clinicians to use a variety of imaging modalities in their diagnostic and treatment options. Each imaging modality has its advantages and disadvantages and must be used in conjunction with clinical judgment.

REFERENCES

1. Jarvik JG, Deyo RA. Diagnostic evaluation of low back pain with emphasis on imaging. *Ann Intern Med.* 2002;137:586-597.
2. Robbins SE, Morse MH. Is the acquisition of a separate view of the sacroiliac joints in the prone position justified in patients with back pain? *Clin Radiol.* 1996;51:637-638.

3. Gran JT. An epidemiological survey of the signs and symptoms of ankylosing spondylitis. *Clin Rheumatol.* 1985;4:161-169.
4. Kalichman L, Kim DH, Li L, Guermazi A, Hunter DJ. Computed tomography-evaluated features of spinal degeneration: prevalence, intercorrelation, and association with self-reported low back pain. *Spine J.* 2010;10:200-208.
5. Resnick R, Niwayama G. Degenerative disease of the spine. In: Resnick D, ed. *Diagnosis of Bone and Joint Disorders.* Philadelphia, PA: WB Saunders Co; 1995:1372-1462.
6. Carrera GF, Haughton VM, Syvertsen A, Williams AL. Computed tomography of the lumbar facet joints. *Radiology.* 1980;134:145-148.
7. Haughton V. Imaging techniques in intraspinal diseases. In: Resnick D. ed. *Diagnosis of Bone and Joint Disorders.* Philadelphia, PA: WB Saunders Co; 1995:237-276.
8. Haughton V. Imaging intervertebral disc degeneration. *J Bone Joint Surg Am.* 2006;88(suppl 2):15-20.
9. Pfirrmann CW, Metzdorf A, Zanetti M, Hodler J, Boos N. Magnetic resonance classification of lumbar intervertebral disc degeneration. *Spine (Phila Pa 1976).* 2001;26:1873-1878.
10. Thompson JP, Pearce RH, Schechter MT, Adams ME, Tsang IK, Bishop PB. Preliminary evaluation of a scheme for grading the gross morphology of the human intervertebral disc. *Spine (Phila Pa 1976).* 1990;15:411-415.
11. Modic MT, Masaryk TJ, Mulopulos GP, Bundschuh C, Han JS, Bohlman H. Cervical radiculopathy: prospective evaluation with surface coil MR imaging, CT with metrizamide, and metrizamide myelography. *Radiology.* 1986;161:753-759.
12. Kettler A, Wilke HJ. Review of existing grading systems for cervical or lumbar disc and facet joint degeneration. *Eur Spine J.* 2006;15:705-718.
13. Fardon DF, Milette PC. Nomenclature and classification of lumbar disc pathology. Recommendations of the Combined Task Forces of the North American Spine Society, American Society of Spine Radiology, and American Society of Neuroradiology. *Spine (Phila Pa 1976).* 2001;26:E93-E113.
14. Thornbury JR, Fryback DG, Turski PA, et al. Disk-caused nerve compression in patients with acute low-back pain: diagnosis with MR, CT myelography, and plain CT. *Radiology.* 1993;186:731-738.
15. Song KJ, Choi BW, Kim GH, Kim JR. Clinical usefulness of CT-myelogram comparing with the MRI in degenerative cervical spinal disorders: is CTM still useful for primary diagnostic tool? *J Spinal Disord Tech.* 2009;22:353-357.
16. Nguyen-minh C, Haughton VM, Papke RA, An H, Censky SC. Measuring diffusion of solutes into intervertebral disks with MR imaging and paramagnetic contrast medium. *AJNR Am J Neuroradiol.* 1998;19:1781-1784.
17. Karnaze MG, Gado MH, Sartor KJ, Hodges FJ III. Comparison of MR and CT myelography in imaging the cervical and thoracic spine. *AJR Am J Roentgenol.* 1988;150:397-403.
18. Shafaie FF, Wippold FJ II, Gado M, Pilgram TK, Riew KD. Comparison of computed tomography myelography and magnetic resonance imaging in the evaluation of cervical spondylotic myelopathy and radiculopathy. *Spine (Phila Pa 1976).* 1999;24:1781-1785.

4

TRAUMATIC CONDITIONS OF THE SPINE

Christopher Loo, MD, PhD; Harvey E. Smith, MD;
and Bradley K. Weiner, MD

EPIDEMIOLOGY

Approximately 30 million injuries necessitating medical care occur annually in the United States. Cervical spine injuries occur in 2% to 4% of all trauma patients, with nearly 30,000 neck injuries documented annually. Injuries to the thoracic and lumbar spine account for more than 50% of all spinal fractures and a large portion of acute spinal cord injuries.[1] Among these, approximately 10,000 cervical spine fractures and 4000 thoracolumbar spine fractures are diagnosed.[2]

Rihn JA, Harris EB. *Musculoskeletal Examination of the Spine: Making the Complex Simple* (pp. 50-61).
© 2011 SLACK Incorporated.

The cause of traumatic spinal cord injury in individuals <65 years include motorcycle accidents, motor vehicle accidents, pedestrian motor vehicle accidents, falls from heights >10 feet, and gunshot wounds. Among the elderly, especially those >75 years, lower energy impacts, such as falls from seated or standing heights, are the most common cause of clinically unstable spine injuries.[2]

TRAUMA WORK-UP

According to *Advanced Trauma and Life Support* guidelines as outlined by the American College of Surgeons,[3] the following list outlines the approach to a patient in the acute trauma setting:

1. Immobilization
2. Films and imaging
   a. Plain radiographs
   b. Computed tomography (CT)
   c. Magnetic resonance imaging (MRI)
3. Complete neurologic examination focusing on the presence of point tenderness, neurologic deficits, or both

The following is a list of critical questions to be addressed during the work-up:

1. What is the patient's Glasgow Coma Scale score?
2. What was the mechanism of injury?
3. Are other distracting injuries present?
4. Is there any focal vertebral body tenderness?
5. Are there any neurologic deficits?
6. Is there any indication the patient was under the influence of alcohol or drugs?

High-risk patients include those with vertebral point tenderness, a Glasgow Coma Score <8, a high-energy mechanism of injury, altered sensorium (eg, trauma, intoxication, or medications), other distracting injuries, a fall from a height >10 feet, and the confirmed presence of other spinal fractures. Patients in the high-risk category have a high probability of spinal injury. The presence of any of these risk factors is an indication for imaging. In contrast, low-risk patients (eg, ambulatory patients without midline tenderness, patients who are able to attain a sitting position, victims of simple rear-end motor vehicle crashes, and those who can actively turn their heads 45 degrees in both directions) were deemed not to require imaging, according to National Emergency X-Radiography Utilization Study criteria.[4,5] There is 100% sensitivity when using these criteria for imaging trauma patients.[6]

The following is a list of questions to be addressed during the imaging work-up:

1. Who needs cervical spine radiographs in the emergency department?

2. When should flexion-extension radiographs be obtained?

3. What are the indications for CT scans, MRI studies, or fluoroscopic evaluation?

4. In the absence of bony injury, how is significant soft-tissue injury to the cervical spine excluded from the differential diagnosis?

ROLE OF RADIOGRAPHS

Standard radiographs have traditionally been the accepted method for screening for cervical fractures. Although the standard number of views is not agreed on, ordering a standard 3 views (ie, anteroposterior, lateral, and odontoid views) carries a 93% sensitivity rate.[6] Lateral views should include the occipitocervical and cervicothoracic junctions. Odontoid views will indicate injury to the body, facet joints, and odontoid process of C2 (Figure 4-1). Some clinicians advocate ordering 5 views, which includes right and left oblique views in addition to the standard 3 views, to look for injuries that are harder to detect.

Standard radiographs are good for screening occipital condyle fractures, Jefferson fractures, atlantoaxial rotatory dislocations, burst fractures with retropulsed fragments, and injuries to the cervicothoracic junction.[7,8] Bagley[6] showed that the main drawbacks such as missed injuries, delayed diagnosis, persistent pain, and permanent neurologic sequelae to

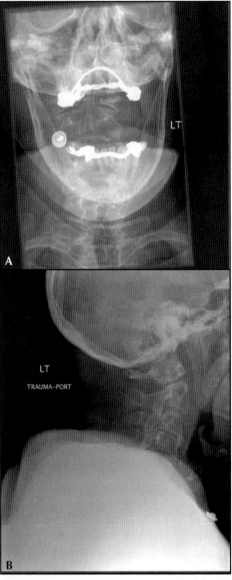

Figure 4-1. (A and B) Open-mouth and lateral radiographs of the cervical spine demonstrating Type II odontoid fracture. Due to patient body habitus, the lower cervical spine and cervicothoracic junction is inadequately visualized, necessitating CT. *(continued)*

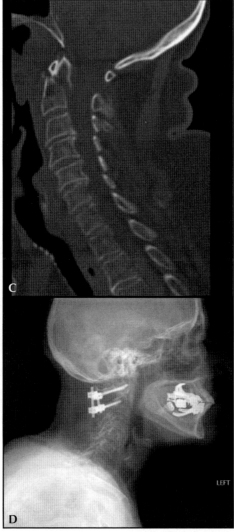

Figure 4-1 (continued). (C) Evaluation of the cervicothoracic junction. (D) Postoperative image showing C1-2 fusion.

standard radiographs were primarily due to poor radiographic quality and misinterpretation by clinicians.

The use of flexion-extension radiographs are not universally agreed on in the acute trauma setting. Due to the significant risk of spinal instability and injury to the spinal column, these should not be obtained in the high-risk trauma patient. However, it is generally agreed on that in a low-risk patient (ie, a fully coherent patient, capable of sitting upright, with normal radiographs) complaining of neck pain flexion-extension views may be ordered to rule out any ligamentous injury.[9]

ROLE OF COMPUTED TOMOGRAPHY

The advent of multi-detector CT scanners has dramatically decreased the amount of time needed to screen for vertebral column injuries.[10] In addition, CT scanners have greatly increased the speed and accuracy of the reconstructions.[11] The sensitivity of CT in detecting cervical spine injuries ranges from 90% to 99%, with specificities ranging from 72% to 89%.[9] Numerous studies comparing standard radiographs to CT have shown CT to be far superior to detecting lesions either undetectable or missed by standard radiography.[10,11]

Given the high probability of missing injuries, delayed diagnosis, and risk for permanent neurologic injury in the moderate-to high-risk trauma victim, CT is rapidly becoming the initial screening modality for spine trauma in adults.[6,9,12] However, in a polytrauma patient with hemodynamic instability, a lateral view of the cervical spine is still commonly obtained because a well-performed lateral cervical spine radiograph with visualization from the occipitocervical junction through the cervicothoracic junction can provide enough information to allow the trauma patient to proceed to the operating room without additional intervention aside from the maintenance of a collar.[13] Thoracic and lumbar injuries are detected by sagittal and coronal reconstructions from the multi-detector CT scans of the chest, abdomen, and pelvis. There is increasing recognition of the radiation exposure of CT scans (Figure 4-2).[6]

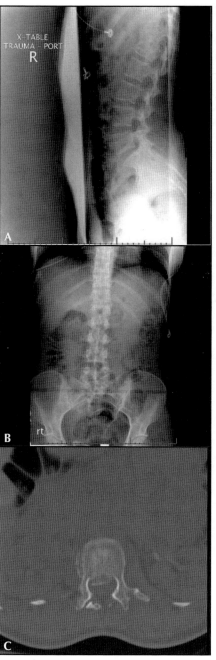

Figure 4-2. (A and B) Lateral and AP radiographs showing a lumbar fracture. (C) CT. *(continued)*

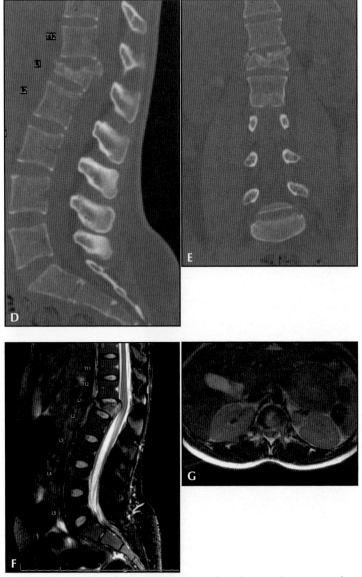

Figure 4-2 (continued). (D and E) Sagittal and coronal views confirm burst component with canal compromise. (F and G) T2-weighted MRIs illustrate the degree of neural element compression.

ROLE OF MAGNETIC RESONANCE IMAGING

MRI is used to evaluate soft tissues, spinal cord, disks, and ligamentous structures. T1-weighted images are used to evaluate spinal cord integrity, T2-weighted images are used to look for the presence of hemorrhage or edema, and axial T2-weighted images are used to look for spinal cord conformation. Fat-suppressed T2-weighted images are used to look at the musculoskeletal components, including the ligamentous structures. In the presence of normal-appearing radiographs, fat-suppressed T1-weighted and T2-weighted sagittal images are typically ordered to evaluate the integrity of the posterior ligamentous complex.[14]

MRI is more sensitive at detecting soft-tissue, disk, ligamentous, and spinal lesions[15] but is less sensitive at detecting bony lesions.[6] MRI is sometimes recommended first in the patient with obvious neurologic deficits. An MRI should be obtained within 48 hours.[15] MRI should be used in conjunction with a good physical examination because of the high false-positive rate (25%)[16] (see Figure 4-2).

CLEARING THE CERVICAL SPINE

Clearance of the cervical spine remains an important issue in the management of the patient with spine trauma. Unnecessary prolonged immobilization in a rigid cervical collar is uncomfortable for the patient and increases the risk of pressure sores and ulcers.[17,18] However, there is no consensus for a comprehensive protocol.[19]

This is particularly relevant in the assessment of ligamentous injuries, as well as in patients with altered sensorium due to head injury or alcohol and drug intoxication. These patients are unable to provide the necessary clinical feedback and thus are at increased risk for complications due to missed injuries.[20]

Missed or delayed diagnosis of a cervical spine injury can lead to the onset of a neurologic injury or allow an incomplete lesion to progress.[21,22] The chance for missed injuries is much greater in patients who present with polytrauma, exhibit drug or alcohol intoxication, are nonresponsive, and have distraction injuries.[21,22] Therefore, a standardized, reliable, and validated approach among clinicians is necessary.

In the fully cooperative patient with no vertebral tenderness, no distracting injuries, full range of motion, and no neurologic injury, routine radiographs are not necessary.[4,5] However, uncertainty arises in the following list of clinical scenarios:

1. The patient with negative radiographs but persistent neck pain, neurologic deficits, or both

2. The polytrauma patient

3. The obtunded patient

In the patient with neck pain but no neurologic deficits and negative radiographs, several options exist. The first option is to obtain MRI to rule out the presence of any ligamentous or spinal cord injury. The second option is to use flexion-extension radiographs or to have the patient actively flex and extend his or her neck. However, problems with patient splinting may invalidate the study. In this case, it is advised that the patient keep the cervical collar on until follow-up, at which point the patient will exhibit clinical improvement or further imaging studies can be performed.[9]

In the obtunded patient, the clinician must weigh the risk of prolonged collar use versus the risk of injury associated with further radiographic intervention. Traditionally, the use of passive fluoroscopic dynamic flexion-extension views or MRI has been used in this scenario. Harris et al[13] recommended the use of a stretch test prior to initiating fluoroscopic-guided flexion-extension views. However, the risk of spinal cord injury due to cervical instability in a patient unable to provide clinical feedback has led some to abandon this recommendation.[21]

The risks of injuring an obtunded patient with possible cervical instability or a critical patient during transport, as well as issues with patient stability in a nonacute clinical setting, such as in a radiography or MRI suite, have led some to recommend several different options[22]:

1. Remove the collar immediately.

2. Keep the collar in place and obtain advanced imaging and a comprehensive clinical examination with the patient is more awake.

3. Remove the cervical collar after a period of time.

A study by Dunham et al[23] showed a higher rate of complications with prolonged cervical collar use as well as greater risks associated with obtaining an MRI (ie, transportation, using an MRI machine in a noncritical care setting, positioning) versus removal of the cervical collar immediately (2.5% risk of cervical instability versus greater risks associated with further radiographic intervention).

CONCLUSION

Imaging in spinal trauma encompasses a variety of modalities including plain radiographs, CT, and MRI, as well as adjunctive imaging modalities. The specific protocols used in the acute trauma setting depend on a variety of factors including the clinical status of the patient, risk factors, and institutional policies. The clinician must maintain a high index of suspicion throughout the evaluation and proceed with a systematic and thorough evaluation. The mechanism of injury and the findings of the clinical examination provide information that will weight the index of suspicion and guide the use of selective imaging modalities. One must exclude not only bone injury but also the possibility of ligamentous instability.

REFERENCES

1. US Department of Health and Human Services. *National Hospital Ambulatory Medical Care Survey, 1994.* Washington, DC: US Department of Health and Human Services; 1994.
2. Mann FA, Cohen WA, Linnau KF, Hallam DK, Blackmore CC. Evidence-based approach to using CT in spinal trauma. *Eur J Radiol.* 2003;48(1): 39-48.
3. American College of Surgeons, Committee on Trauma. *Advanced Trauma Life Support.* Chicago, IL: American College of Surgeons; 1997.
4. Hoffman JR, Wolfson AB, Todd K, Mower WR. Selective cervical spine radiography in blunt trauma: methodology of the National Emergency X-Radiography Utilization Study (NEXUS). *Ann Emerg Med.* 1998;32(4): 461-469.
5. Viccellio P, Simon H, Pressman BD, Shah MN, Mower WR, Hoffman JR. A prospective multicenter study of cervical spine injury in children. *Pediatrics.* 2001;108(2):E20.
6. Bagley LJ. Imaging of spinal trauma. *Radiol Clin North Am.* 2006;44(1): 1-12.

7. Streitwieser DR, Knopp R, Wales LR, Williams JL, Tonnemacher K. Accuracy of standard radiographic views in detecting cervical spine fractures. *Ann Emerg Med*. 1983;12(9):538-542.

8. Pech P, Kilgore DP, Pojunas KW, Haughton VM. Cervical spinal fractures: CT detection. *Radiology*. 1985;157(1):117-120.

9. el-Khoury GY, Kathol MH, Daniel WW. Imaging of acute injuries of the cervical spine: value of plain radiography, CT, and MR imaging. *AJR Am J Roentgenol*. 1995;164(1):43-50.

10. Lynch D, McManus F, Ennis JT. Computed tomography in spinal trauma. *Clin Radiol*. 1986;37(1):71-76.

11. France JC. Update on the appropriate radiographic studies for cervical spine: evaluation and clearance in the polytraumatized patient. *Curr Orthop Pract*. 2008;19(4):411-415.

12. Geusens E, Van Breuseghem I, Pans S, Brys R. Some tips and tricks in reading cervical spine radiographs in trauma patients. *JBR-BTR*. 2005;88(2):87-92.

13. Harris MB, Kronlage SC, Carboni PA, et al. Evaluation of the cervical spine in the polytrauma patient. *Spine (Phila Pa 1976)*. 2000;25(22): 2884-2891.

14. Lee HM, Kim HS, Kim DJ, Suk KS, Park JO, Kim NH. Reliability of magnetic resonance imaging in detecting posterior ligament complex injury in thoracolumbar spinal fractures. *Spine (Phila Pa 1976)*. 2000;25(16): 2079-2084.

15. Cohen WA, Giauque AP, Hallam DK, Linnau KF, Mann FA. Evidence-based approach to use of MR imaging in acute spinal trauma. *Eur J Radiol*. 2003;48(1):49-60.

16. Petersilge CA, Pathria MN, Emery SE, Masaryk TJ. Thoracolumbar burst fractures: evaluation with MR imaging. *Radiology*. 1995;194(1):49-54.

17. Plaisier B, Gabram SG, Schwartz RJ, Jacobs LM. Prospective evaluation of craniofacial pressure in four different cervical orthoses. *J Trauma*. 1994;37(5):714-720.

18. Chendrasekhar A, Moorman DW, Timberlake GA. An evaluation of the effects of semirigid cervical collars in patients with severe closed head injury. *Am Surg*. 1998;64(7):604-606.

19. Reid DC, Henderson R, Saboe L, Miller JD. Etiology and clinical course of missed spine fractures. *J Trauma*. 1987;27(9):980-986.

20. Stiell IG, Wells GA, Vandemheen K, et al. Variation in emergency department use of cervical spine radiography for alert, stable trauma patients. *CMAJ*. 1997;156(11):1537-1544.

21. Bednar DA, Toorani B, Denkers M, Abdelbary H. Assessment of stability of the cervical spine in blunt trauma patients: review of the literature, with presentation and preliminary results of a modified traction test protocol. *Can J Surg*. 2004;47(5):338-342.

22. Sliker CW, Mirvis SE, Shanmuganathan K. Assessing cervical spine stability in obtunded blunt trauma patients: review of medical literature. *Radiology*. 2005;234(3):733-739.

23. Dunham CM, Brocker BP, Collier BD, Gemmel DJ. Risks associated with magnetic resonance imaging and cervical collar in comatose, blunt trauma patients with negative comprehensive cervical spine computed tomography and no apparent spinal deficit. *Crit Care*. 2008;12(4):R89.

III

Common Conditions of the Spine

5

CERVICAL DISK HERNIATION

Alan S. Hilibrand, MD and Kris Radcliff, MD

INTRODUCTION

Cervical disk herniation is a common condition that results when the inner component of the disk, the nucleus pulposus, protrudes into or extrudes through an incompetent annulus fibrosus. Herniation is an anatomical or radiographic finding. Musculoskeletal examination is critical to the accurate diagnosis of clinical syndromes resulting from symptomatic cervical disk herniation.

Cervical disk herniation is most common is the third and fourth decades of life. In a landmark study, Boden et al determined that 10% of asymptomatic patients under 40 have cervical disk herniation on MRI and 5% of asymptomatic patients

Rihn JA, Harris EB. *Musculoskeletal Examination of the Spine: Making the Complex Simple* (pp. 63-81).
© 2011 SLACK Incorporated.

over 40 have disk herniation.[1] Posterior disk protrusion was observed in over 70% of subjects in a prospective, natural history study of ageing in the cervical spine, while only 10% and 4% were noted to have developed new radiculopathy symptoms during the study period.[2] The most commonly affected level affected are C5-6, C4-5, and C6-7, in that order.[3]

Cervical disk herniation may be asymptomatic, cause neck pain alone, or produce neurological symptoms of radiculopathy, and/or myelopathy. Radiculopathy is caused when a cervical disk herniation causes inflammation and/or compression of a spinal nerve root exiting the spinal cord. Cervical myelopathy describes a constellation of symptoms and physical findings which are the result of spinal cord dysfunction. There are other possible anatomical causes of radiculopathy and myelopathy, including osteophytes, trauma, fractures, tumors, infections/abscess, and congenital conditions. However, cervical disk herniations are the most common cause of cervical radiculopathy and myelopathy in young patients.

The diagnosis of cervical disk herniation is fairly straightforward with modern imaging techniques, including MRI. The treatment of cervical disc herniation ranges from conservative management (eg, physical therapy, traction, anti-inflammatory medication, epidural injections) to surgical management. The purpose of this chapter is to review the presentation, diagnosis, and management of cervical disc herniation.

HISTORY

Patients with a symptomatic cervical disc herniation typically present with neck pain, arm pain, arm/hand numbness or tingling, and/or arm weakness. The typical cervical disc herniation is posterolateral (ie, off to one side), thus impinging on the nerve root that exits the spinal canal at the level of the herniation. The nerve root involved is typically the lower numbered nerve root at any given cervical level. For example, a patient with a right paracentral C6-7 disc herniation will have compression of the right C7 nerve root.

The precise pain generator from cervical disk herniation is thought to be a combination of chemical and inflammatory mediators around the nerve root and direct pressure on the nerve root. The result of these events caused by the disc herniation can cause arm symptoms.[4-6] Radiating arm pain is the most common complaint in patients with cervical radiculopathy.[7] Sensory symptoms, predominantly parasthesias and numbness, are also common. Loss of strength may also be reported by patients with cervical radiculopathy. The arm symptoms typically follow the distribution of the cervical nerve root that is compressed and/or inflamed in a dermatomal or myotomal distriubtion. The experience of the authors is that the myotomal distribution of pain is more reliable and more easily assessed than the dermatomal distribution. Radicular symptoms are often exacerbated by maneuvers that stretch or impinge the involved nerve root, such as turning the head during driving, coughing, sneezing, Valsalva, neck extension, and certain cervical movements and positions. Patients occasionally prefer shoulder abduction and may rest the wrist and hand on top of the head for relief of the arm symptoms.[8] The seventh cervical nerve root is the most frequently involved by cervical radiculopathy.[9]

Large, central cervical disc herniations that compress the spinal cord can cause myelopathy. The clinical definition of myelopathy is deterioration of coordinated movements of the upper and/or lower extremities which usually occur in the presence of pathologic reflexes such as a Hoffman's sign, Babinski's sign, or myoclonus of the ankle. These long-tract findings are the result of disinhibition of the reflex arc of the spinal afferent or efferent nerve tracts.[10,11] Myelopathy often has an insidious onset. When examining a patient with a cervical disc herniation, it is necessary to directly inquire about the symptoms of myelopathy as patients may not associate them or bring them to the examiner's attention. Myelopathy typically presents with a loss of manual dexterity, such as difficulty with fine motor movements, buttoning buttons, penmanship, typing.[12,13] Myelopathy also results in loss of proprioception, ataxia, and bowel and bladder dysfunction. Bowel and bladder dysfunction may include overflow incontinence, urge incontinence, or frank incontinence.

Attributing axial neck pain to cervical disc herniation or degeneration is a source of controversy because of the high number of asymptomatic degenerative changes in the cervical disks. However, the cervical disks are innervated, particularly in the posterior annulus where most herniations occur.

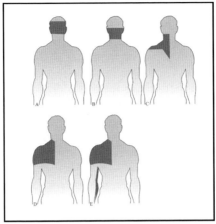

Figure 5-1. Referred dermatomes of cervical diskogenic pain.

Cervical discography has identified patterns of pain produced by stimulation of each cervical disk, although there is a high overlap between dermatomes (Figure 5-1).[14]

Standard questions in the patient history include the age of the patient, the nature of the complaint, the anatomic distribution, the timing of the complaint, possible inciting events, ameliorating and exacerbating conditions, previous treatments and response, and a careful review of symptoms. The authors also asks the patients to draw the distribution of radiculopathy on their body, particularly distal to the forearm.

The other crucial element of the history is identification of red flags that suggest a more complex condition such as night pain, pain unrelated to activity or position, a history of IV drug abuse or malignancy, constitutional symptoms, weight change, cranial nerve symptoms, or symptoms in a nonanatomic distribution. The clinician should directly inquire the patients about their work status, their home relationships, the presence of any active compensation claims, and the presence of any active or pending litigation. Associated conditions must be ruled out with questions directed at possible peripheral nerve entrapment such as carpal tunnel or cubital tunnel syndrome, possible shoulder involvement, and neurologic conditions especially multiple sclerosis. Although it is outside of the scope of this chapter, the author inquires about night hand pain or morning hand numbness and hand numbness when gripping a steering wheel of an automobile as screening questions for carpal tunnel syndrome. Similarly, the author will inquire about positional or temporal distribution of cubital tunnel syndrome. Weakness outside of proportion to clinical pain complaint is a red flag for consideration of amyotrophic lateral sclerosis or some other neurological condition. Signs of lower (LMN) and upper motor neuron (UMN) loss in multiple body regions (bulbar, cervical, thoracic and/or lumbosacral) are diagnostic criteria for amyotrophic lateral sclerosis. Approximately 10-15% of ALS patients undergo possibly unnecessary surgery prior for an incorrect diagnosis.[15]

EXAMINATION

Examination for cervical disk herniation should involve the entire musculoskeletal system, not only the upper extremities (Table 5-1). Observation commences with gait, tandem gait, heel walk, and toe walk. Myelopathy will result in a slow, broad-based, shuffling gait and inability to tandem walk. Observation of the range of motion of the cervical spine should include an evaluation of flexion, extension, lateral bending, and rotation. The clinician should screen for asymmetry, especially in lateral bending and rotation, which typically narrows the ipsilateral neuroforamen and exacerbates the symptoms of a disk herniation. The examiner should carefully test light-touch and pinprick sensation in all dermatomes of the upper and lower extremity. During cervical extension, the author will also inquire about symptoms of Lhermitte's sign, a lightening sensation over the distribution of the entire spine.

In the case of radiculopathy, the sensory examination is initially focused on determination of the symptomatic dermatome of examination. Strength testing of muscles according to myotomes is the next tested element. The examiner should use his or her knowledge of dermatomes and myotomes to attempt to differentiate peripheral neuropathy from cervical radiculopathy. Finally, reflexes should be assessed, including biceps, brachioradialis, triceps, patellar, and Achilles reflexes. A rectal examination and test of perianal sensation is indicated for a suspicion of possible incontinence.

Table 5-1

METHODS OF EXAMINATION: CERVICAL DISK HERNIATIONS

Nerve Level	Disk Level	Myotome	Reflex	Dermatome
C3	C2-C3	Trapezius, levator scapulae, strap muscles, sternocleidomastoid, and diaphragm	None	Suboccipital and posterior auricular regions
C4	C3-C4	Trapezius, rhomboids, levator scapulae, and diaphragm	None	Posterorlateral cervical regions and posterior shoulder
C5	C4-C5	Pectoralis major (clavicular head), supraspinatus, infraspinatus, deltoid, biceps, brachialis, brachioradialis, and diaphragm	None	Deltoid area
C6	C5-C6	Biceps, brachialis, brachioradialis, extensor carpi radialis longus, supinator, pronator teres, flexor carpi radialis, and triceps	Biceps and brachioradialis	Lateral arm, forearm, thumb, and index finger
C7	C6-C7	Triceps, latissimus dorsi, pronator teres, flexor carpi radialis, extensor carpi ulnaris, extensor digitorum, abductor pollicis longus, extensor pollicis brevis and longus, and extensor indicis	Triceps	Posterior shoulder, posterior arm, dorsal forearm, and middle finger

(continued)

Table 5-1 (continued)

METHODS OF EXAMINATION: CERVICAL DISK HERNIATIONS

Nerve Level	Disk Level	Myotome	Reflex	Dermatome
C8	C7-T1	Flexor digitorum superficialis, pronator quadratus, flexor digitorum profundus, flexor pollicis longus, flexor carpi ulnaris, and lumbricals 3 and 4	None	Medial arm, medial forearm, ring and small digits
T1	T1-T2	Adductor pollicis, abductor pollicis brevis, opponens pollicis, flexor pollicis brevis, interossei, and lumbricals 1 and 2; Horner syndrome may be present	None	Axillary and pectoral region, medial arm and proximal medial forearm

Adapted from Abbed K, Coumans JV. Cervical radiculopathy. *Neurosurgery.* 2007;60(1)(suppl 1):S28-S34.

Specialized provocative testing is the next phase of the cervical spine physical examination. Spurling's sign is produced with a combination of extension, lateral rotation, and gentle axial loading to compress the neuroforamina on the side of the rotation. The test is positive when it produces ipsilateral arm pain. The sensitivity and specificity of Spurling's test for cervical radiculopathy are 30% and 93%, respectively.[16] Similarly, shoulder abduction may decrease stress within the nerve root and relieve radicular pain. This shoulder abduction sign may be helpful to differentiate C8 radiculopathy versus thoracic outlet syndrome.

To rule out intracranial pathology, the examiner should perform a cranial nerve examination, evaluate for tongue fasciculations, and consider a jaw jerk reflex, which consists of reflex closing of the mouth with a downward tap on the jaw with the mouth open.[17,18] Additionally, Romberg's sign, which involves the patient maintaining forearm supination with arms extended and eyes closed for 45 seconds to 1 minute, is helpful to identify cerebellar pathology as a source of poor coordination. Any consideration of intracranial pathology should prompt referral to a neurologist for more in-depth testing.

Myelopathy should be sought on the physical exam of all spine patients. Some myelopathic signs include hyperreflexia of the deep tendon reflexes of the lower extremities in the setting of hypo- or hyper-reflexia in the upper extremities, increased muscle tone or clonus, and the presence of pathological reflexes resulting from disinhibition.[19-21] Babinski's sign is slow dorsiflexion of the MTP joint of the great toe associated with peripheral stimulation of the plantar foot. Hoffman's sign consists of flexion of the interphalangeal joint of the thumb with extension of the DIP joint of the long finger. A recent study identified that Hoffman's sign is present in 81% of patients with severe and 46% of patients with mild myelopathy. Babinski's sign is present in 10% of patients with mild compression and 83% of severe compression patients.[22] Previous studies have identified that Hoffman's sign can be present in individuals with asymptomatic spinal cord compression.[23] Inverted radial reflex is contraction of the index finger flexors with stimulation of the brachioradialis reflex. Lhermitte's sign is a sign of general spinal cord pathology. Severe muscle atrophy caudal to the level of stenosis is uncommon with a spondylotic myelopathy. Overall, approximately

79% of patients with cervical myelopathy would be expected to have physical signs, as compared to 57% of control patients.[24]

Any question of shoulder pathology should merit a specific examination including shoulder range of motion and provocative tests for impingement and bicipital tendonitis. Similarly, a concern for peripheral neuropathy should include an evaluation of carpal or cubital tunnel specific tests, including Durkin's test, Tinel's sign, or Phalen's sign. A list of differential diagnoses for specific cervical radiculopathy levels is included in Table 5-2.

Pathoanatomy

The outer portion of the disk is made up of the annulus fibrosus. Ventrally, it is multilaminated with interweaving fibers of alternating orientation, but dorsally, it is only present as a thin layer of collagen fibers.[25] The mechanically weakest portion of the annulus is posterolateral and the majority of disk herniations occur in this location. Disk herniations can also occur centrally. Cervical disk herniations usually create radiculopathy in the nerve root corresponding to the caudal level. For example, the C6 nerve root exits above the C6 pedicle and is affected by a C56 disc herniation. This differs from the thoracolumbar spine where the numbered nerve root exits caudal to the numbered pedicle.

Disk herniations can be "soft," meaning that they contain nucleus pulposus. A "hard" disk herniation represents spondylosis and may contain osteophytes (ie, bone spurs). Other details of pathoanatomy include whether the herniation is a protruded annular bulge (ie, the nucleus pushes against a thinned annulus but does not go through it), extruded (ie, the nucleus is completely herniated through the annulus), or sequestered (ie, the nucleus is completely herniated through the annulus and separated from the disk all together). Disk herniations may be resorbed and remodeled by the body either partially or completely.

Radiculopathy results from a combination of inflammation and mechanical distortion of the nerve root by protruding annulus, herniated nucleus pulposus, or uncovertebral osteophytes. The exact pathogenesis of radicular pain is unclear and may involve mechanical, inflammatory, or vascular etiologies.

Table 5-2

HELPFUL HINTS: DIFFERENTIAL DIAGNOSIS OF CERVICAL DISK HERNIATIONS

Nerve Level	Entity Mimicking Radiculopathy	Differentiating Factor
C3	Suboccipital headaches	Exacerbated by neck motion if spine etiology; Spurling's sign may be present
C5	Shoulder pathology, including frozen shoulder or rotator cuff tear	Presence of impingement signs and presence of pain with passive shoulder motion, relieved by abduction if radiculopathy and exacerbated if shoulder impingement
C6	Median nerve entrapment, carpal tunnel syndrome, and cubital tunnel syndrome	Provocative tests for carpal tunnel syndrome, including Durkin test, Phalen test, and Tinel sign; thenar atrophy is specific for carpal tunnel syndrome as the thenar muscles are T1 innervated; forearm symptoms suggest either pronator syndrome or radiculopathy
C7	Carpal tunnel syndrome if volar pain; posterior interosseous nerve (PIN) syndrome if wrist extensor weakness; radial tunnel syndrome if dorsal and painful	PIN syndrome is not a pain syndrome and usually presents with isolated weakness of PIN innervated muscles; brachioradialis and extensor carpi radialis longus are radial nerve, not PIN innervated and should help differentiate; radial tunnel syndrome is rare; associated with wrist extension
C8-T1	Ulnar nerve entrapment	Presence of Tinel sign or Durkin test should exacerbate cubital tunnel syndrome; isolated C8 entrapment should not affect the hand interossei and thus the Froment paper sign should be negative
	Thoracic outlet syndrome	Thoracic outlet syndrome is exacerbated by shoulder abduction

Adapted from Abbed K, Coumans JV. Cervical radiculopathy. Neurosurgery. 2007;60(1)(suppl 1):S28-S34.

Cervical spondylotic myelopathy is the manifestation of long tract signs resulting from a decrease in the space available for the cervical spinal cord from mechanical compression. Myelopathy may also be related to the intrinsic anteroposterior diameter of the spinal canal, dynamic cord compression, dynamic changes in the intrinsic morphology of the spinal cord, and the vascular supply of the spinal cord.

IMAGING

Cervical radiographs are most useful in identifying associated conditions, such as spondylosis, spondylolisthesis, fractures, dislocations, tumors, or infection. An assessment of mechanical cervical alignment is helpful for preoperative planning.[26] Like MRI, radiographs are also frequently abnormal in asymptomatic individuals.[27]

MRI may also enable evaluation of inflammatory, neoplastic, degenerative, or pathologic changes in the spinal cord (Figures 5-2 and 5-3). If the patient is unable to undergo MRI, then a CT myelogram offers excellent visualization of neural compression although it is an invasive study. CT myelogram is also excellent for evaluation of associated spondylosis conditions such as OPLL. Occasionally a clinician may consider both MRI and CT as preoperative studies to rule out the presence of spondylosis prior to a procedure, such as a posterior foraminotomy and diskectomy, that is contingent upon the presence of a soft disk herniation.

The timing of MRI is controversial. MRI should be delayed until symptoms of radiculopathy have been present for 6 weeks in the absence of warning signs such as constitutional symptoms, night pain, pain unrelated to activity or position, history of intravenous drug abuse or malignancy, fevers, weight change, cranial nerve symptoms, symptoms in a nonanatomic distribution, bowel or bladder dysfunction, and significant or progressive neurologic deficit. MRI should be ordered more urgently for possible diagnosis of myelopathy, tumors, and degenerative neurologic conditions. MRI is highly sensitive for pathology but is less specific.

It is important to note that degenerative findings and disk herniations found on MRI do not necessarily correlate to

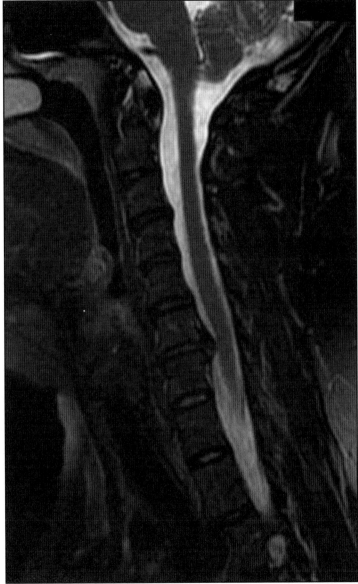

Figure 5-2. Midsagittal T2-weighted MRI shows a cervical disk herniation at C6-C7 effacing the spinal cord.

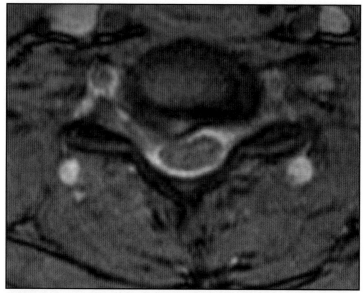

Figure 5-3. Transaxial T2-weighted MRI at C6-C7 shows a large right-sided disk herniation.

symptoms. In one classic study, 8% of asymptomatic patients were noted to have disk herniations. Degenerative disease may be observed in 25% of asymptomatic people less than 40 years of age and 60% of those older than 40.[1]

Electrodiagnostic testing (ie, electromyelography/nerve conduction study) can be helpful in the evaluation of patients with a symptomatic cervical disk herniation. Electrodiagnostic examination can accurately determined the correct nerve root level in 57% of the patients in a study comparing EMG and intraoperative findings.[28]

TREATMENT

The natural history of cervical radiculopathy has been demonstrated to be generally favorable.[29-32] At long-term follow-up (2 to 19 years) of 51 patients with radiculopathy, 45% had only a single episode of pain without recurrence, 30% had mild symptoms, and only 25% had persistent or worsening symptoms. No radiculopathy progressed to myelopathy. The benign natural history justifies a trial of at least 6 weeks of nonoperative treatment. Treatment of cervical radiculopathy may include a trial of anti-inflammatory medications, physical therapy, traction, and corticosteroid injections. There are few studies comparing nonsurgical treatment methods. Transforaminal and interlaminar cervical injections have been described.[33-36] Transforaminal injections carry a risk of intravascular injection which can result in stroke and death.[35] Interlaminar injection can result in spinal cord injury.[36] Most clinicians will attempt a trial of nonoperative treatment for at least 6 weeks prior to consideration of surgical intervention for radiculopathy, given the possibility of resorption of cervical disk herniations.

Surgery is indicated early for progressive worsening symptoms of weakness, incontinence, fractures, tumors, or other signs of neurological deterioration. Additionally, if a patients has persistent symptoms after nonsurgical treatment of radiculopathy, then surgery may be considered. The most common treatment of cervical disk herniation is anterior cervical discectomy and fusion (Figure 5-4). This approach affords a wide exposure of the neural elements and a wide decompression and has a high success rate historically.[37-42] However, this approach has associated morbidities of dysphagia and dysphonia[43] and controversy exists as to whether the associated fusion may increase degeneration at other segments.[44] Posterior approaches are also successful for soft foraminal disc herniations and radiculopathy, particularly far lateral soft disc herniations.[45] The outcome of diskectomy for posterior soft disk herniations is equivalent to anterior cervical surgery. Additionally, although it is a nonfusion procedure, the rate of adjacent segment disease after posterior cervical surgery is equivalent to that after anterior operations.[46] Soft cervical disc herniations lateral to the spinal cord are amenable to posterior cervical foraminotomies because they require minimal retraction of the cervical spinal cord and nerve roots. Cervical disc replacement is a new treatment option for disc herniations (Figure 5-5).

Some patients with cervical myelopathy will not progress. A trial of nonoperative treatment is indicated for the management of mild or early myelopathy in the absence of persistent radiculopathy and does not appear to impair the ultimate outcome of the patients.[47-54] The natural history of moderate to severe myelopathy is stepwise deterioration in 75% of patients.[55-58] The goal of surgery for myelopathy is halting

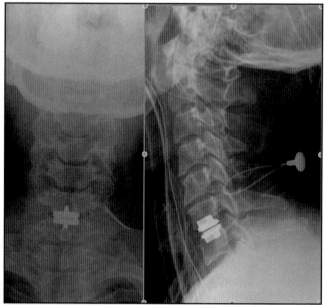

Figure 5-4. Postoperative AP and lateral radiographs show anterior cervical diskectomy and fusion with allograft plating.

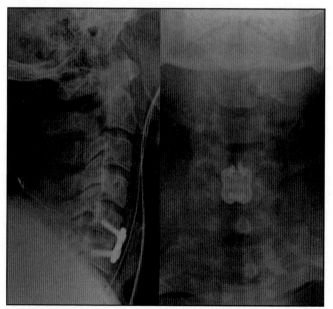

Figure 5-5. Postoperative AP and lateral radiographs show cervical total disk arthroplasty for disk herniation.

progression of the disease. Surgery for myelopathy from disk herniation usually consists of anterior cervical discectomy and fusion to address the anterior pathology.

Axial neck pain from cervical disc herniations may be managed conservatively with physical therapy, traction, and nonsteroidal anti-inflammatory medications. Provocative tests, including discography, may be helpful to directly diagnose the symptomatic level of degeneration although it is rarely used due to the associated risks of visceral/vascular injury as well as the reluctance to treat axial pain through surgical intervention. Fusion for axial pain is generally considered less successful than treatment of radiculopathy.

CONCLUSION

Cervical disc herniation can cause neck pain, radiculopathy, or myelopathy. The most common presenting symptoms are radiculopathy, numbness, paresthesias, or weakness. The care of a patient with cervical myelopathy must involve detailed history, careful physical examination, and judicious use and interpretation of imaging. Decision making is based on knowledge of the natural history of these conditions. A majority of patients with cervical disc herniation and radiculopathy will improve with nonoperative treatment. Those who do not respond to conservative treatment after a minimum of 6 weeks should be considered for surgical treatment. Surgical treatment is typically performed earlier in patients with significant myelopathy.

REFERENCES

1. Boden SD, McCowin PR, Davis DO, Dina TS, Mark AS, Wiesel S. Abnormal magnetic-resonance scans of the cervical spine in asymptomatic subjects. A prospective investigation. *J Bone Joint Surg Am.* 1990;72(8):1178-1184.
2. Okada E, Matsumoto M, Ichihara D, et al. Aging of the cervical spine in healthy volunteers: a 10-year longitudinal magnetic resonance imaging study. *Spine (Phila Pa 1976).* 2009;34(7):706-712.
3. Matsumoto M, Fujimura Y, Suzuki N, et al. MRI of cervical intervertebral discs in asymptomatic subjects. *J Bone Joint Surg Br.* 1998;80(1):19-24.
4. Lee SJ, Han TR, Hyun JK, Jeon JY, Myong NH. Electromyographic findings in nucleus pulposus-induced radiculopathy in the rat. *Spine (Phila Pa 1976).* 2006;31(18):2053-2058.

5. Kawakami M, Hashizume H, Nishi H, Matsumoto T, Tamaki T, Kuribayashi K. Comparison of neuropathic pain induced by the application of normal and mechanically compressed nucleus pulposus to lumbar nerve roots in the rat. *J Orthop Res.* 2003;21(3):535-539.

6. Kawakami M, Tamaki T, Weinstein JN, Hashizume H, Nishi H, Meller ST. Pathomechanism of pain-related behavior produced by allografts of intervertebral disc in the rat. *Spine (Phila Pa 1976).* 1996;21(18):2101-2107.

7. Rao R. Neck pain, cervical radiculopathy, and cervical myelopathy: pathophysiology, natural history, and clinical evaluation. *Instr Course Lect.* 2003;52:479-488.

8. Davidson RI, Dunn EJ, Metzmaker JN. The shoulder abduction test in the diagnosis of radicular pain in cervical extradural compressive monoradiculopathies. *Spine (Phila Pa 1976).* 1981;6(5):441-446.

9. Rao R. Neck pain, cervical radiculopathy, and cervical myelopathy: pathophysiology, natural history, and clinical evaluation. *J Bone Joint Surg Am.* 2002;84-A(10):1872-1881.

10. Harrop JS, Naroji S, Maltenfort M, et al. Cervical myelopathy: a clinical and radiographic evaluation and correlation to cervical spondylotic myelopathy [published online ahead of print February 10, 2010]. *Spine (Phila Pa 1976).* doi: 10.1097/BRS.0b013e3181b723af

11. Harrop JS, Hanna A, Silva MT, Sharan A. Neurological manifestations of cervical spondylosis: an overview of signs, symptoms, and pathophysiology. *Neurosurgery.* 2007;60(1 Suppl 1):S14-S20.

12. Spillane JD, Lloyd GH. Spastic paraplegia in late adult life; with degeneration and protrusion of cervical discs. *Lancet.* 1951;2(6685):653-657.

13. Spillane JD, Lloyd GH. The diagnosis of lesions of the spinal cord in association with osteoarthritic disease of the cervical spine. *Brain.* 1952;75(2):177-186.

14. Grubb SA, Kelly CK. Cervical discography: clinical implications from 12 years of experience. *Spine (Phila Pa 1976).* 2000;25(11):1382-1389.

15. Bedlack RS. Amyotrophic lateral sclerosis: current practice and future treatments. *Curr Opin Neurol.* 2010;23(5):524-529.

16. Tong HC, Haig AJ, Yamakawa K. The Spurling test and cervical radiculopathy. *Spine (Phila Pa 1976).* 2002;27(2):156-159.

17. de Watteville A. A description of the cerebral and spinal nerves of Rana Esculenta. *J Anat Physiol.* 1874;9(Pt 1):145-162.

18. de Watteville A. On the tendon-reactions. *Br Med J.* 1886;1(1329):1160.

19. Bohlman HH, Emery SE. The pathophysiology of cervical spondylosis and myelopathy. *Spine (Phila Pa 1976).* 1988;13(7):843-846.

20. Emery SE. Cervical spondylotic myelopathy: diagnosis and treatment. *J Am Acad Orthop Surg.* 2001;9(6):376-388.

21. Emery SE. Surgical management of cervical myelopathy. *Instr Course Lect.* 1999;48:423-426.

22. Houten JK, Noce LA. Clinical correlations of cervical myelopathy and the Hoffmann sign. *J Neurosurg Spine.* 2008;9(3):237-242.

23. Sung RD, Wang JC. Correlation between a positive Hoffmann's reflex and cervical pathology in asymptomatic individuals. *Spine (Phila Pa 1976).* 2001;26(1):67-70.

24. Rhee JM, Heflin JA, Hamasaki T, Freedman B. Prevalence of physical signs in cervical myelopathy: a prospective, controlled study. *Spine (Phila Pa 1976).* 2009;34(9):890-895.

25. Mercer S, Bogduk N. The ligaments and annulus fibrosus of human adult cervical intervertebral discs. *Spine (Phila Pa 1976)*. 1999;24(7):619-626; discussion 627-618.

26. Dean CL, Gabriel JP, Cassinelli EH, Bolesta MJ, Bohlman HH. Degenerative spondylolisthesis of the cervical spine: analysis of 58 patients treated with anterior cervical decompression and fusion. *Spine J*. 2009;9(6):439-446.

27. Gore DR. Roentgenographic findings in the cervical spine in asymptomatic persons: a ten-year follow-up. *Spine (Phila Pa 1976)*. 2001;26(22): 2463-2466.

28. Partanen J, Partanen K, Oikarinen H, Niemitukia L, Hernesniemi J. Preoperative electroneuromyography and myelography in cervical root compression. *Electromyogr Clin Neurophysiol*. 1991;31(1):21-26.

29. Gore DR, Sepic SB, Gardner GM, Murray MP. Neck pain: a long-term follow-up of 205 patients. *Spine (Phila Pa 1976)*. 1987;12(1):1-5.

30. Lees F. Cervical Spondylosis. *Nurs Times*. 1964;60:1240-1242.

31. Lees F, Turner JW. Natural history and prognosis of cervical spondylosis. *Br Med J*. 1963;2(5373):1607-1610.

32. Chrispin AR, Lees F. The spinal canal in cervical spondylosis. *J Neurol Neurosurg Psychiatry*. 1963;26:166-170.

33. Smuck M, Yu AJ, Tang CT, Zemper E. Influence of needle type on the incidence of intravascular injection during transforaminal epidural injections: a comparison of short-bevel and long-bevel needles. *Spine J*. 2010;10(5):367-371.

34. Smuck M, Rosenberg JM, Akuthota V. The use of epidural corticosteroids for cervical radiculopathy: an interlaminar versus transforaminal approach. *PM R*. 2009;1(2):178-184.

35. Smuck M, Fuller BJ, Chiodo A, et al. Accuracy of intermittent fluoroscopy to detect intravascular injection during transforaminal epidural injections. *Spine (Phila Pa 1976)*. 2008;33(7):E205-E210.

36. Strub WM, Brown TA, Ying J, Hoffmann M, Ernst RJ, Bulas RV. Translaminar cervical epidural steroid injection: short-term results and factors influencing outcome. *J Vasc Interv Radiol*. 2007;18(9):1151-1155.

37. Bohlman HH, Emery SE, Goodfellow DB, Jones PK. Robinson anterior cervical discectomy and arthrodesis for cervical radiculopathy. Long-term follow-up of one hundred and twenty-two patients. *J Bone Joint Surg Am*. 1993;75(9):1298-1307.

38. Bohlman HH, Anderson PA. Anterior decompression and arthrodesis of the cervical spine: long-term motor improvement. Part I--improvement in incomplete traumatic quadriparesis. *J Bone Joint Surg Am*. 1992;74(5):671-682.

39. Emery SE, Fisher JR, Bohlman HH. Three-level anterior cervical discectomy and fusion: radiographic and clinical results. *Spine (Phila Pa 1976)*. 1997;22(22):2622-2624; discussion 2625.

40. Emery SE, Bohlman HH, Bolesta MJ, Jones PK. Anterior cervical decompression and arthrodesis for the treatment of cervical spondylotic myelopathy. Two to seventeen-year follow-up. *J Bone Joint Surg Am*. 1998;80(7):941-951.

41. Hilibrand AS, Yoo JU, Carlson GD, Bohlman HH. The success of anterior cervical arthrodesis adjacent to a previous fusion. *Spine (Phila Pa 1976)*. 1997;22(14):1574-1579.

42. Gore DR, Sepic SB. Anterior cervical fusion for degenerated or protruded discs. A review of one hundred forty-six patients. *Spine (Phila Pa 1976).* 1984;9(7):667-671.
43. Rihn JA, Kane J, Albert TJ, Vaccaro AR, Hilibrand AS. What is the incidence and severity of dysphagia after anterior cervical surgery [published online ahead of print December 8, 2010]? *Clin Orthop Relat Res.* doi: 10.1007/s11999-010-1731-8.
44. Hilibrand AS, Carlson GD, Palumbo MA, Jones PK, Bohlman HH. Radiculopathy and myelopathy at segments adjacent to the site of a previous anterior cervical arthrodesis. *J Bone Joint Surg Am.* 1999;81(4): 519-528.
45. Henderson CM, Hennessy RG, Shuey HM, Jr., Shackelford EG. Posterior-lateral foraminotomy as an exclusive operative technique for cervical radiculopathy: a review of 846 consecutively operated cases. *Neurosurgery.* 1983;13(5):504-512.
46. Herkowitz HN, Kurz LT, Overholt DP. Surgical management of cervical soft disc herniation. A comparison between the anterior and posterior approach. *Spine (Phila Pa 1976).* 1990;15(10):1026-1030.
47. Kadanka Z, Mares M, Bednarik J, et al. Predictive factors for spondylotic cervical myelopathy treated conservatively or surgically. *Eur J Neurol.* 2005;12(1):55-63.
48. Kadanka Z, Mares M, Bednarik J, et al. Predictive factors for mild forms of spondylotic cervical myelopathy treated conservatively or surgically. *Eur J Neurol.* 2005;12(1):16-24.
49. Bednarik J, Sladkova D, Kadanka Z, et al. Are subjects with spondylotic cervical cord encroachment at increased risk of cervical spinal cord injury after minor trauma [published online ahead of print June 28, 2010]? *J Neurol Neurosurg Psychiatry.* Jun 28 2010. doi:10.1136/jnnp.2009.198945.
50. Bednarik J, Kadanka Z, Dusek L, et al. Presymptomatic spondylotic cervical myelopathy: an updated predictive model. *Eur Spine J.* 2008;17(3):421-431.
51. Bednarik J, Kadanka Z, Dusek L, et al. Presymptomatic spondylotic cervical cord compression. *Spine (Phila Pa 1976).* 2004;29(20):2260-2269.
52. Kadanka Z, Bednarik J, Vohanka S, Stejskal L, Smrcka V, Vlach O. Spondylotic cervical myelopathy: three aspects of the problem. *Suppl Clin Neurophysiol.* 2000;53:409-418.
53. Kadanka Z, Mares M, Bednanik J, et al. Approaches to spondylotic cervical myelopathy: conservative versus surgical results in a 3-year follow-up study. *Spine (Phila Pa 1976).* 2002;27(20):2205-2210; discussion 2210-2201.
54. Kadanka Z, Bednarik J, Vohanka S, et al. Conservative treatment versus surgery in spondylotic cervical myelopathy: a prospective randomised study. *Eur Spine J.* 2000;9(6):538-544.
55. Clarke E. Cervical myelopathy presenting as peripheral neuropathy. *Br Med J.* 1958;1(5082):1282-1284.
56. Clarke E, Robinson PK. Cervical myelopathy: a complication of cervical spondylosis. *Brain.* 1956;79(3):483-510.
57. Clarke E, Little JH. Cervical myelopathy; a contribution to its pathogenesis. *Neurology.* 1955;5(12):861-867.
58. Clarke E. Cervical myelopathy; a common neurological disorder. *Lancet.* 1955;268(6856):171-176.

6

CERVICAL
SPONDYLOSIS

David T. Anderson, MD and Todd J. Albert, MD

INTRODUCTION

Cervical spondylosis is a common condition that includes a continuum of degenerative, age-related changes resulting in compression and inflammation of the nerve roots (radiculopathy), spinal cord (myelopathy), or a combination of the two (myeloradiculopathy). A progression of events leads to changes in the intervertebral disks, vertebral bodies, facet joints, and ligaments of the cervical spine. The end result of compression and inflammation of the neural elements has multiple causal factors. Direct compression due to osteophytes, disk herniation, ossified ligaments, or folded ligaments is one

Rihn JA, Harris EB. *Musculoskeletal Examination of the Spine: Making the Complex Simple* (pp. 82-97).
© 2011 SLACK Incorporated.

Table 6-1

HELPFUL HINTS: PRIMARY CERVICAL SPONDYLOSIS SYNDROMES

Spondylotic Syndromes	Description/Mechanism	Typical Patient Complaints
Neck pain (cervicalgia)	Disk (diskogenic pain) or facet degeneration	Axial neck pain, often worse with motion
Radiculopathy	Nerve root compression due to herniated disk, foraminal stenosis, or osteophyte formation	Radiating arm pain, paresthesias, and weakness
Myelopathy	Spinal cord compression due to central disk herniation, osteophytes, and folded ligamentum flavum	Loss of fine motor skills, gait dysfunction, stiffness of legs, bowel/bladder dysfunction

common variable. Other contributing factors include vascular insufficiency and venous engorgement. In addition, congenital differences in anatomy, including underlying congenital stenosis, may change the ability of the neural elements to respond to injury. As this is a natural occurrence with aging, it is often difficult to distinguish pathologic changes from normal physiologic changes. Therefore, anatomic changes on imaging must be carefully correlated with clinical symptoms and signs to ensure proper management.

Cervical spondylosis encompasses 3 primary syndromes (Table 6-1), with multiple overlapping qualities, often seen in conjunction with one another:

- Cervicalgia (nonradiating neck pain)
- Radiculopathy
- Myelopathy

Neck pain can be acute or chronic and often is the result of disk degeneration (ie, diskogenic pain). Radiculopathy is a pathologic process resulting from compression and inflammation of a cervical nerve root. It may present as weakness, sensory changes, or radiating pain in a dermatomal distribution. Myelopathy results from compression of the spinal cord itself and presents in only a fraction of patients with spondylotic changes.[1] It commonly consists of gait disturbance, difficulty with fine motor skills, spastic weakness and numbness in the hands, and long tract findings.[1]

Cervical spondylosis is common in industrial populations; 10% of individuals have some element of the disease by age 25 and 95% by age 65.[2] Although earlier reports suggest the course of the disease results in a progressive neurologic decline,[3,4] more recent investigations note that for the majority of cases with mild symptoms, there is an initial phase of decline followed by a static period of stable symptoms.[5] However, these reports also note that older patients deteriorate more frequently and that surgery should be reserved for older patients with progressive disability. Pain reduction and strengthening are the goals of conservative treatment.[6] Surgery is indicated for radiculopathy if conserve treatment fails or if there is profound or progressive weakness or intractable pain. Surgery for myelopathy is indicated if there is significant signs and symptoms including balance disturbance, bowel or bladder dysfunction, or problems with coordination that correlate with magnetic resonance imaging (MRI) evidence of spinal cord compression. Finally, other conditions mimicking radiculopathy (eg, primary shoulder disease, upper-extremity nerve entrapment, or peripheral neuropathy) and myelopathy (eg, multiple sclerosis, amyotrophic lateral sclerosis, tumors, normal pressure hydrocephalus, or epidural abscess) must be ruled out.

HISTORY

After symptoms of cervical spondylosis develop, they generally fall into the 3 primary syndromes of neck pain, radiculopathy, and myelopathy (see Table 6-1). Neck pain often presents without a distinctive precipitating event and is often attributed to disk degeneration.[2] Degenerative facet joints can also cause neck pain. Pain charts are often used for predicting the segmental location of symptomatic joints in patients with cervical pain.[7,8] This can help the examiner differentiate referred joint pain from the classic distribution of radiculopathy.

Radiculopathy traditionally presents in the dermatomal distribution and can be unilateral, bilateral, symmetric, or

Table 6-2

MOTOR GRADING

5	Full strength against resistance
4	Less than full strength against resistance
3	Able to lift against gravity
2	Able to move with gravity eliminated
1	Muscle twitch
0	No twitch

asymmetric. Patients <55 years usually have symptoms attributed to a herniated disk, whereas the symptoms of those 55 and older can usually be attributed to spondylotic changes, such as foraminal stenosis due to osteophyte formation.[9] Weakness and atrophy are often associated with a soft disk herniation, whereas sensory changes are more commonly associated with hard disk degeneration. Sensory changes include paresthesias, hyperesthesias, and hyperalgesias. In more chronic conditions, reflex changes may become apparent.[10]

Cervical myelopathy may be progressive or may follow a course in which there is an initial deterioration followed by a stable period.[5] It may manifest as a loss of fine motor skills, such as fastening buttons or changes in handwriting, along with gait dysfunction or stiffness of the legs.[4] Neck stiffness and axial pain can also be present. Rarely, loss of sphincter control or urinary symptoms occurs. Acutely, a central cord syndrome may be observed in a patient with clinically quiet cervical spondylotic myelopathy after a traumatic hyperextension.

PHYSICAL EXAMINATION

A complete neurologic examination must be performed in the evaluation of suspected cervical spondylosis. Motor strength (Tables 6-2 and 6-3), sensation to light touch and

Table 6-3

CERVICAL MOTOR STRENGTH TESTING

Muscle Group	Nerve/Root Level	Technique
Deltoid	Axillary/C5	Resisted shoulder abduction
Brachialis/biceps	Musculocutaneous/ C5-C6	Resisted elbow flexion
Wrist extensors (ECRL, ECRB)	Radial/C6	Resisted wrist extension
Finger extensors (EDC, EIP, EDM)	Radial (PIN)/C7	Resisted finger extension
Finger flexors	Median/C8	Resisted finger/thumb flexion
Hand intrinsics	Ulnar/T1	Resisted finger abduction/ adduction, finger cross

ECRB indicates extensor carpi radialis brevis; ECRL, extensor carpi radialis longus; EDC, extensor digitorum communis; EDM, extensor digiti minimus; EIP, extensor indicis proprius; PIN, posterior interosseous nerve.

Table 6-4

REFLEXES

Nerve Root	Muscular Reflex
C5	Biceps
C6	Brachioradialis
C7	Triceps

pinprick, muscular reflexes (Table 6-4), and reflexes of specific long tracts should be assessed. In examining the patient with suspected radiculopathy, in addition to the motor strength and sensation examination, the following tests may help in defining pathology (Table 6-5).

Table 6-5

SPECIFIC TESTS

Examination	Technique	Grading/Results	Significance
Shoulder abduction test	Passive abduction of patient's affected arm	Relief of symptoms	Indicates radiculopathy
Spurling's test	Extend neck, rotate toward side with symptoms, and axial load	Radiating pain	Specific for radiculopathy
Lhermitte's sign	Flexion of neck with gentle axial load	Electric shock-like sensation down the spine	Indicates posterior column dysfunction
Hoffmann's sign	Forceful flexion of the distal phalanx of the third digit, followed by a sudden release	Resultant flexion and abduction of the thumb with simultaneous flexion of the index finger is positive	Upper motor neuron dysfunction
Babinski's reflex	Blunt object scraped along the lateral border of the plantar aspect of the foot from heel to ball and then curved medially across the metatarsophalangeal joints	Slow tonic dorsiflexion of the great toe with fanning of the lesser toes	Upper motor neuron dysfunction
Inverted radial reflex	Gentle tapping of the brachioradialis tendon	Hyperactive finger flexion	Upper motor neuron dysfunction
Finger escape sign	Patient holds his or her fingers extended and adducted	Abduction of ulnar digits within 30 to 60 seconds is positive	Upper motor neuron dysfunction

Shoulder Abduction Test

In the shoulder abduction test, as described by Davidson et al,[11] the patient's radicular symptoms should be relieved when the shoulder is abducted. In theory, this maneuver releases tension on the nerve root.

Spurling's Test

In Spurling's test, pain is elicited by having the patient extend and rotate his or her neck toward the side with pain while the examiner applies axial load, thereby narrowing the neural foramen. Myelopathy is defined by the presence of long tract findings. Hyperreflexia of deep tendon reflexes, increased muscle tone, clonus, and the presence of pathologic reflexes are the classic signs.

Babinski's Reflex

Babinski's reflex is elicited with a blunt object scraped along the lateral border of the plantar aspect of the foot from heel to ball and then curved medially across the metatarsophalangeal joints. A slow tonic dorsiflexion of the great toe with fanning of the lesser toes is considered Babinski's sign.[12]

Hoffmann's Sign

Hoffmann's sign is the upper-extremity equivalent of Babinski's sign. It is elicited by stimulating the extensor tendon of the third digit with forceful flexion of the distal phalanx, followed by a sudden release. If there is a resultant flexion and abduction of the thumb with simultaneous flexion of the index finger, then the test is considered positive.[13] Sung and Wang[13] suggest that "the presence of a positive Hoffmann's reflex in asymptomatic patients strongly suggests underlying cervical pathology, but it does not warrant further evaluation with either cervical radiographs or magnetic resonance imaging since the management and clinical course are not affected by positive studies."

Lhermitte's Sign

Another useful sign is Lhermitte's sign, which consists of an electric shock-like sensation shooting down the spine with flexion of the neck. This is thought to be the result of posterior column dysfunction.

Additional Signs

Additional signs include the inverted radial reflex and the finger escape sign. By stimulating the distal brachioradialis tendon with gentle tapping, hyperactive finger flexion is considered a positive inverted radial reflex. The finger escape sign is evaluated by having the patient hold his or her fingers extended and adducted. If the ulnar digits drift into abduction within 30 to 60 seconds, the test is considered positive.

In determining the prevalence and utility of commonly tested myelopathic signs, Rhee et al[14] showed that hyperreflexia or provocative signs (ie, Hoffmann's, inverted radialis reflex, clonus, and Babinski's) are absent in up to 20% of patients with cervical myelopathy. They concluded that symptoms combined with correlative imaging studies must be used to base treatment decisions and that absence of signs does not preclude a diagnosis nor does it predict successful surgical treatment.

PATHOANATOMY OF CERVICAL SPONDYLOTIC MYELOPATHY/ RADICULOPATHY

In cervical spondylosis, primary degenerative processes are responsible for the ultimate secondary compressive phenomena. Both neural and vascular processes can become symptomatic. White and Panjabi[15] developed a theory of the biomechanical factors leading to cervical spodylotic myeloradiculopathy. They divided factors into static and dynamic groups. The static mechanisms relate to the primary degenerative processes that lead to a smaller canal diameter. These would include a congenitally small canal, osteophyte formation from the vertebral body or uncovertebral joint, disk herniation, hypertrophy of the facet joint, ligamentum flavum folding, and calcification of the posterior longitudinal ligament or ligamentum flavum. The dynamic factors include abnormal forces on the spinal cord during loading and movement. Together, these mechanisms can lead to neurologic symptoms.

Reduction in the sagittal spinal canal size is the driving force behind development of spondylotic symptoms. This process begins with cervical disk desiccation. The cervical disk is taller ventrally than it is dorsally, which effectively maintains lordosis. The ventral anulus fibrosus is made of Type I collagen interweaved and multilayered, whereas the dorsal annulus is merely a thin layer of fibers.[16] As the body ages, there is a progressive decline in the water content of the intervertebral disk. The nucleus pulposus is made of glycosaminoglycans, a proteoglycan core with many polysaccharide attachments of keratan sulfate and chondroitin sulfate, negatively charged to hold water. This unit gives the nucleus its elastic and viscous properties. With aging, the nucleus pulposus shows morphologically high cell activities near the cartilaginous end plate, including a process suggestive of regeneration. Eventually it becomes an indistinct fibrocartilaginous mass.[17] As a result, the disk losses its ability to retain water, its elasticity, and its ability to bear load. With no other options, the annulus fibrosus takes on the weight-bearing duties.

As this degenerative process ensues, the contents of the disk can bulge into the spinal canal.[1,18] Initially, the loss of height occurs ventrally and may lead to a loss of lordosis. Altered biomechanics lead to increased forces ventrally and may eventually lead to a kyphotic deformity.[18] As the vertebral bodies subside toward one another, the facet joint capsule and the ligamentum flavum folds into the spinal canal, thus further reducing the canal diameter. Due to increased load bearing, osteophytes are formed at the edges of the vertebral body, facet joints, and uncovertebral joints.

All of these effects of aging eventually lead to reduced spinal canal and encroachment on the neural elements. White and Panjabi[15] discovered that patients with a sagittal canal diameter <14.8 mm were more likely to develop cervical spondylotic myelopathy. The location of the compressive pathology (spinal cord, nerve root, or a combination of both) determines the spectrum of symptoms. Compression between the uncovertebral joint and the facet joint may cause radicular symptoms, whereas compression from vertebral body osteophytes, bulging disks, ossified posterior longitudinal ligament, or infolded ligamentum flavum may cause myelopathy.

Cervical motion also affects the secondary compressive processes. Flexion may exacerbate or cause symptoms from ventral compression. Extension may induce compression from infolding of the ligamentum flavum. In addition, unstable cervical segments may result in a pincer phenomenon, during which flexion or extension may pinch the spinal cord between ventral and dorsal structures.[15] Another biomechanical consideration is the "sagittal bowstring effect," in which the cord is tethered over a ventral mass in a kyphotic spine. In this situation, a posterior laminectomy may actually worsen the deformity, and an anterior approach to the pathology is warranted.[18]

Not only do the compressive processes affect the spinal cord itself, they also may compress local blood supply to the cord, resulting in ischemia. Both arterial and venous vessels can be affected by the degenerative changes in the cervical spine. Necrosis and cavitation may be seen in the spinal gray matter, which seems to be more sensitive to ischemic events. The spinal cord has a more tenuous radicular blood supply in the C5 to C7 region.[19] In addition, it has been thought that acute presentation of cervical myelopathy may be due to the thrombosis of a compressed artery.[10,20,21]

At the cellular and molecular level, recent literature has shed light on the pathobiological mechanisms associated with the progressive loss of neural tissue in cervical spondylotic myelopathy. Yu et al[22] used an animal model consisting of the twy/twy mutant mouse, which develops an ossified ligamentum flavum at C2-C3 and undergoes progressive paralysis, to show that chronic cord compression leads to Fas-mediated apoptosis of neurons and oligodendrocytes. This, in turn, is associated with activation of caspase-8, caspase-9, and caspase-3, and progressive neurologic deficits. They suggest that molecular therapies eventually could complement surgical decompression to help maximize neurologic recovery.

Free radical- and cation-mediated cell injury and glutamertergic toxicity may also play a role in the pathophysiology of cervical myelopathy.[23] It seems that there is early demyelination of the corticospinal tracts followed by destruction of the anterior horn cells. There is relative preservation of the anterior columns. Changes generally occur caudal to compressive lesions. Necrosis and cavitation in the central gray matter are seen late and suggest chronic disease.[24]

IMAGING

Radiographs

Despite the widespread availability and utility of more advanced imaging, plain radiographs still hold their place in the comprehensive clinical evaluation of cervical disease. Radiographs remain an important screening tool, are inexpensive, and are easily obtained. They can provide information regarding sagittal balance, deformity, fractures, congenital abnormalities, and instability. Flexion-extension radiographs can uncover instability in segments that may be a cause of motion-induced pain.

Computed Tomography

Now readily available and quite fast, computed tomography (CT) allows direct visualization of the bony elements causing compression of the neural structures. CT provides greater detail in the neural foramina than MRI.[25] CT has high spatial resolution and allows the surgeon to understand the pathology in 3 dimensions. It is a useful adjunct to MRI and plain radiographs in surgical planning.

Magnetic Resonance Imaging

MRI can display soft-tissue anatomy precisely and noninvasively. It has become the most common and useful imaging modality for suspected cervical spine pathology. It is important to note that MRI often shows disease that is subclinical. Ernst et al[26] studied the MRIs of 30 asymptomatic volunteers. They found bulging disks in 73% and focal disk protrusions in 50%. Thirty-seven percent had annular tears at one or more levels. Degenerative disk disease may be found in 25% of asymptomatic patients <40 years and in 60% of those 40 and older.[27] For these reasons, MRI must be obtained judiciously and closely correlated with symptoms and the clinical examination. Nevertheless, MRI is a key component in the evaluation for suspected cervical pathology.

TREATMENT

Although still debated, it is generally agreed that the natural history of the majority of patients with cervical spondylotic myelopathy involves an early phase of deterioration that is followed by a static period lasting for several years, in which there is little change in the level of disability.[1,5] Nurick[5] noted that the majority of patients who declined rapidly were older and recommended surgery be reserved for those with more advanced progressive disease who were >60 years. Patients with moderate to severe disability were more likely to do well with surgery, and patients with mild symptoms were not likely to worsen.[28] Therefore, the consensus has been to surgically treat those patients with documented disability given the tendency for myelopathy to gradually progress. Other authors have supported the role of early intervention even for mild symptoms because patients with cervical spondylotic myelopathy may be at higher risk for serious spinal cord injury due to minor trauma.[29,30]

Conservative treatment includes strengthening, physical therapy, and pain management. Selective nerve root injection can relieve radicular symptoms while simultaneously confirming level of pathology. These conservative measures are typically performed for patients with radicular symptoms rather than myelopathy. Cervical interlaminar epidural injections are typically not performed in myelopathic patients due to the risk of worsening the neurologic status of the patient. Surgical treatment rather than conservative treatment is typically recommended for patients with significant myelopathy.

Decompression of the spinal cord or nerve roots and stabilization of the spinal column are the ultimate goals of surgery. Most often the anterior compressive pathology is located at the disk space(s), in the form of a disk/osteophyte complex. Anterior decompression usually involves removing the diseased disk, placing a structural bone graft (either allograft or iliac crest autograft) into the debrided disk space, and stabilizing the level using a plate-and-screw construct (ie, anterior cervical diskectomy and fusion) (Figures 6-1 and 6-2).

If the compressive pathology extends behind the vertebral body itself, then a partial or complete corpectomy (ie, removal of the vertebral body) must be performed. The structural bone graft required for a corpectomy is larger. A multilevel corpectomy requires a long structural graft anteriorly and often requires a combined posterior approach as well to place posterior screws and rods for added stability. A posterior alone decompression (ie, cervical laminectomy or laminoplasty) can be used to decompress the spinal cord when the compression exists at multiple levels. To perform a posterior decompression, the cervical spine must have some lordosis, thus allowing the spine to drift away from the anterior pathology (ie, disk/osteophyte complexes) after the laminectomy or laminoplasty is

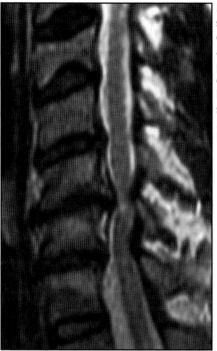

Figure 6-1. Sagittal MRI shows compressive pathology of the cervical spine at 2 levels. The disk/osteophyte complex has impinged on the cord. This patient had worsening myelopathic symptoms and underwent a 2-level anterior cervical diskectomy and fusion.

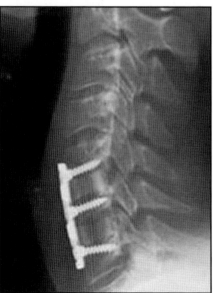

Figure 6-2. Postoperative lateral C-spine radiograph shows the plate-and-screw construct after 2-level anterior cervical diskectomy and fusion. The disk spaces were removed and bone graft can be seen in their place.

Figure 6-3. Sagittal MRI shows multilevel spondylotic changes with a kyphotic deformity. This patient required a combined anterior-posterior procedure.

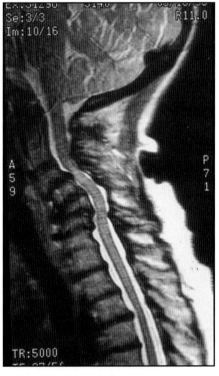

performed. In a kyphotic cervical spine, the spinal cord will not drift away from the anterior pathology after performing a posterior alone decompression (Figure 6-3). For this reason, patients with kyphosis of the cervical spine and multilevel cervical spondylotic myelopathy usually require a combined anterior-posterior surgical approach.

Overall, the natural course of cervical spondylotic myelopathy is unpredictable.[31] However, it does seem that after moderate symptoms have developed, they are less likely to improve on their own and may very well progress, thus suggesting a benefit from surgery.[32]

CONCLUSION

Cervical spondylosis is a common source of disability in the middle-aged and older population. Although its natural course generally involves an initial deterioration followed by a static period, the disease can be rapidly progressive. It is paramount that the spine surgeon understands its pathophysiology, presentation, diagnosis, and possible treatment options. A thorough history and physical examination combined with appropriate imaging should yield an accurate diagnosis and confirm location of pathology.

REFERENCES

1. Asgari S. Cervical spondylotic myelopathy. In: Palmer JD, ed. *Neurosurgery 1996: Manual of Neurosurgery*. New York, NY: Churchill Livingstone; 1996:750-754.
2. Garfin SR. Cervical degenerative disorders: etiology, presentation, and imaging studies. *Instr Course Lect*. 2000;49:335-338.
3. Crandall PH, Batzdorf U. Cervical spondylotic myelopathy. *J Neurosurg*. 1966;25(1):57-66.
4. Spillane JD, Lloyd GH. The diagnosis of lesions of the spinal cord in association with osteoarthritic disease of the cervical spine. *Brain*. 1952;75(2):177-186.
5. Nurick S. The natural history and the results of surgical treatment of the spinal cord disorder associated with cervical spondylosis. *Brain*. 1972;95(1):101-108.
6. Saal JS, Saal JA, Yurth EF. Nonoperative management of herniated cervical intervertebral disc with radiculopathy. *Spine (Phila Pa 1976)*. 1996;21(16):1877-1883.
7. Dwyer A, Aprill C, Bogduk N. Cervical zygapophyseal joint pain patterns, I: a study in normal volunteers. *Spine (Phila Pa 1976)*. 1990;15(6): 453-457.
8. Aprill C, Dwyer A, Bogduk N. Cervical zygapophyseal joint pain patterns, II: a clinical evaluation. *Spine (Phila Pa 1976)*. 1990;15(6):458-461.
9. Truumees E, Herkowitz HN. Cervical spondylotic myelopathy and radiculopathy. *Instr Course Lect*. 2000;49:339-360.
10. Connell MD, Wiesel SW. Natural history and pathogenesis of cervical disk disease. *Orthop Clin North Am*. 1992;23(3):369-380.
11. Davidson RI, Dunn EJ, Metzmaker JN. The shoulder abduction test in the diagnosis of radicular pain in cervical extradural compressive monoradiculopathies. *Spine (Phila Pa 1976)*. 1981;6(5):441-446.
12. Bickley LS, Hoekelman RA. The nervous system. In: Bickley LS, ed. *Bates' Guide to Physical Examination and History Taking*. Baltimore, MD: Lippincott Williams & Wilkins; 1999:595.
13. Sung RD, Wang JC. Correlation between a positive Hoffmann's reflex and cervical pathology in asymptomatic individuals. *Spine (Phila Pa 1976)*. 2001;26(1):67-70.
14. Rhee JM, Heflin JA, Hamasaki T, Freedman B. Prevalence of physical signs in cervical myelopathy: a prospective, controlled study. *Spine (Phila Pa 1976)*. 2009;34(9):890-895..
15. White AA III, Panjabi MM. Biomechanical considerations in the surgical management of cervical spondylotic myelopathy. *Spine (Phila Pa 1976)*. 1988;13(7):856-860.

16. Mercer S, Bogduk N. The ligaments and annulus fibrosus of human adult cervical intervertebral discs. *Spine (Phila Pa 1976)*. 1999;24(7):619-626.

17. Oda J, Tanaka H, Tsuzuki N. Intervertebral disc changes with aging of human cervical vertebra: from the neonate to the eighties. *Spine (Phila Pa 1976)*. 1988;13(11):1205-1211.

18. Benzel ED. *Biomechanics of Spine Stabilization*. Rolling Meadows, IL: American Association of Neurological Surgeons Publications; 2001.

19. Chakravorty BG. Arterial supply of the cervical spinal cord (with special reference to the radicular arteries). *Anat Rec*. 1971;170(3):311-329.

20. Bohlman HH, Emery SE.: The pathophysiology of cervical spondylosis and myelopathy. *Spine (Phila Pa 1976)*. 1988;13(7):843-846.

21. McCormick WE, Steinmetz MP, Benzel ED. Cervical spondylotic myelopathy: make the difficult diagnosis, then refer for surgery. *Cleve Clin J Med*. 2003;70(10):899-904.

22. Yu WR, Baptiste DC, Liu T, Odrobina E, Stanisz GJ, Fehlings MG. Molecular mechanisms of spinal cord dysfunction and cell death in the spinal hyperostotic mouse: implications for the pathophysiology of human cervical spondylotic myelopathy. *Neurobiol Dis*. 2009;33(2): 149-163.

23. Fehlings MG, Skaf G. Spine: a review of the pathophysiology of cervical spondylotic myelopathy with insights for potential novel mechanisms drawn from traumatic spinal cord injury. *Spine (Phila Pa 1976)*. 1998;23(24):2730-2737.

24. Kim RC. Spinal cord pathology. In: Nelson JS, Parisi JE, Schochet SS, eds. *Principles and Practice of Neuropathology*. St. Louis, MO: CV Mosby; 1993:398-435.

25. Freeman TB, Martinez CR. Radiographical evaluation of cervical spondylotic disease: limitation of magnetic resonance imaging for diagnosis and preoperative assessment. *Perspectives in Neurological Surgery*. 1992;3:34-36.

26. Ernst CW, Stadnik TW, Peeters E, Breucq C, Osteaux MJ. Prevalence of annular tears and disc herniations on MR images of the cervical spine in symptom free volunteers. *Eur J Radiol*. 2005;55(3):409-414.

27. Malanga GA. The diagnosis and treatment of cervical radiculopathy. *Med Sci Sports Exerc*. 1997;29(suppl 7):S236-S245.

28. Symon L, Lavender P. The surgical treatment of cervical spondylotic myelopathy. *Neurology*. 1967;17(2):117-127.

29. Firooznia H, Ahn JH, Rafii M, Ragnarsson KT. Sudden quadriplegia after a minor trauma: the role of preexisting spinal stenosis. *Surg Neurol*. 1985;23(2):165-168.

30. Montgomery DM, Brower RS. Cervical spondylotic myelopathy: clinical syndrome and natural history. *Orthop Clin North Am*. 1992;23(3): 487-493.

31. LaRocca H. Cervical spondylotic myelopathy: natural history. *Spine (Phila Pa 1976)*. 1988;13(7):854-855.

32. Emery SE. Cervical spondylotic myelopathy: diagnosis and treatment. *J Am Acad Orthop Surg*. 2001;9(6):376-388.

CERVICAL TRAUMA
UPPER CERVICAL SPINE

Ryan P. Ponton, MD and Eric B. Harris, MD

Upper cervical spine (occiput-C2) trauma can be challenging to a spine surgeon given the often vague and nonspecific presentation, combined with the possibility of deleterious sequelae with these injuries. Any patient where the clinician is considering cervical spine injury must be thoroughly evaluated. Fracture, dislocation, or ligamentous injury of the upper cervical spine should be suspected in any individual who complains of suboccipital headache or pain in the neck area because the patient may be at risk for spinal cord injury unless the spine is stabilized. An important factor contributing to the individual's long-term functional abilities after injury to the upper c-spine is how the injury was managed initially.

Rihn JA, Harris EB. *Musculoskeletal Examination of the Spine: Making the Complex Simple* (pp. 98-124).
© 2011 SLACK Incorporated.

The treatment of a patient with a cervical spine injury is initiated at the scene of the injury. Without exception, all victims of trauma are initially suspected to have a cervical injury until proven otherwise. Because upper c-spine injuries are commonly concomitant with our injuries from trauma, *Advanced Trauma Life Support (ATLS)* protocol should guide initial resuscitation. A protected airway and adequate ventilation are of special importance in injuries of the upper cervical spine because these injuries can lead to diaphragmatic and intercostal paralysis with respiratory failure. In addition, large retropharyngeal hematomas can cause upper airway obstruction. Nasotracheal intubation and cricothyroidotomy are the most expeditious in the acute setting and require less cervical spine motion than direct oral intubation techniques.[1,2] Once the patient's airway, breathing, and circulation are addressed, initial stabilization of the cervical spine may be accomplished with the application of a rigid cervical collar, a spine board, and sandbags.

HISTORY

Given the often nonspecific presentation of upper c-spine injuries, history may provide insight into diagnosis. If the patient is able to give details of the accident, the mechanism of injury (flexion versus extension, compression vs. distraction) should be obtained because typical injury mechanisms can lead to predictable injury patterns.[3] If the individual is unable to communicate, statements from witnesses may prove to be helpful. In addition, episodes of transient neural deficits should also be elicited from either the patient or witnesses, as this may be a sign of an unstable injury.

Upper cervical spine injury has been linked to the presence of severe head injury, high-energy mechanism, and focal neurologic deficits.[4] The most frequent injuries result from motor vehicle collisions, falls, diving into shallow water, and gunshot wounds to the neck.[3]

EXAMINATION

Upper c-spine trauma injuries commonly produce unreliable physical exam findings. All aspects of physical exam: inspection, palpation, neurologic evaluation should be performed with the head and neck stabilized in neutral alignment (Table 7-1). Documentation of the exam should follow the International Standards for Neurological Classification of Spinal Cord Injury (ISCSCI) practice guidelines as published The American Spinal Injury Association (ASIA; Figures 7-1 and 7-2).

With assistance, the patient is log-rolled to one side with head and neck stabilized, so the head and neck can be carefully inspected and palpated. At this time, the collar may be removed and inspected for gross deformity. Observations to document include cranial injuries and facial lacerations, which often occur in conjunction with c-spine injury.[5] Patients may also present with torticollis, so include posture and positioning of the head and neck in your exam. Careful palpation should be performed after inspection. Exam findings to look for and document include areas of tenderness, muscle spasm, step-off or crepitus. It is rare to illicit any other sign other than localized tenderness.

Head and neck range of motion should not be undertaken until screening X-rays are complete. The range of motion between the occiput and atlas is 25 degrees in flexion-extension, 5 degrees to each side in lateral bending, and 5 degrees to each side in rotation. The range of motion between the atlas and axis is 20 degrees in flexion-extension, 5 degrees in lateral bending, and 40 degrees in rotation. Patients can present with normal to wide range of sensorimotor deficits. Depending on level of consciousness, spontaneously movement of all extremities should be documented. Lower cranial nerves, specifically VI, VII, IX, XI, XII, should also be tested for deficits given the course of their fibers. Pathologic (Babinski's and Hoffman's) or irregular (hyperreflexia and clonus) reflexes can be observed with advanced myelopathy.

Table 7-1

METHODS FOR EXAMINING

Examination	Technique	Grading	Notes
Inspection	Log-roll to one side with head and neck stabilized. Collar may be removed at this time to inspect for gross deformity.		Document posture and positioning of the head to evaluate for torticollis. C-spine deformity, cranial injuries, facial lacerations, and facial asymmetries should also be noted.
Palpation	Evaluate areas of tenderness, muscle spasm, step-off or crepitus. It is rare to illicit any other sign other than localized tenderness.		

(continued)

Table 7-1 (continued)

METHODS FOR EXAMINING

Examination		Technique	Grading	Notes
Neurologic	Motor	Evaluate elbow flexors, wrist extensors, elbow extensors, finger flexors, and finger abductors, hip flexors, knee extensors, ankle dorsiflexors, great toe extensors, and ankle plantar flexors.	0 – Total paralysis 1 – Palpable or visible contraction 2 – Active movement, full range of motion, gravity eliminated 3 – Active movement, full range of motion, against gravity 4 – Active movement, full range of motion, against gravity and provides some resistance 5 – Active movement, full range of motion, against gravity and provides normal resistance	Take note of muscle tone (ie, flaccid or spastic)

(continued)

Table 7-1 (continued)

METHODS FOR EXAMINING

Examination		Technique	Grading	Notes
Neurologic	Sensory	Test both light touch and pin prick. See Figure 7-1A for dermatomal distribution.		
	Cranial	Evaluate cranial nerves II-XII		Pay special attention to these cranial nerves given the course of their fibers: VI, VII, IX, XI, XII.
	Reflexes	Brachioradialis, biceps, triceps, patellar tendon, achilles, Hoffman's, and Babinski's reflexes	0: Absent 1+: Hypoactive 2+: Normal 3+: Hyperactive without clonus 4+: Hyperactive with clonus	Pathologic (Babinksi and Hoffman's) or irregular (hyperreflexia and clonus) reflexes can be observed with advanced myelopathy. If reflexes appear hyperactive, test for ankle clonus.

(continued)

Table 7-1 (continued)

METHODS FOR EXAMINING

Examination	Technique	Grading	Notes
Neurologic	ASIA Impairment Scale	A – Complete: No motor or sensory function is preserved in the sacral segments S4-S5.	
		B – Incomplete: Sensory but not motor function is preserved below the neurological level and includes the sacral segments S4-S5.	
		C – Incomplete: Motor function is preserved below the neurological level, and more than half of key muscles below the neurological level have a muscle grade less than 3.	
		D – Incomplete: Motor function is preserved below the neurological level, and at least half of key muscles below the neurological level have a muscle grade of 3 or more.	
		E – Normal: No motor or sensory deficit.	

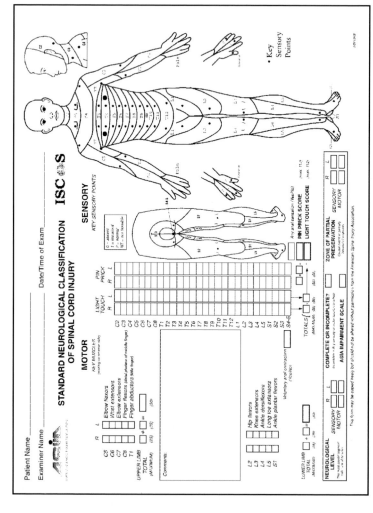

Figure 7-1. International Standards for Neurological Classification of Spinal Cord Injury (ISCSCI) practice guidelines as published by The American Spinal Injury Association (ASIA).

MUSCLE GRADING

0 total paralysis

1 palpable or visible contraction

2 active movement, full range of motion, gravity eliminated

3 active movement, full range of motion, against gravity

4 active movement, full range of motion, against gravity and provides some resistance

5 active movement, full range of motion, against gravity and provides normal resistance

5* muscle able to exert, in examiner's judgement, sufficient resistance to be considered normal if identifiable inhibiting factors were not present

NT not testable. Patient unable to reliably exert effort or muscle unavailable for testing due to factors such as immobilization, pain on effort or contracture.

ASIA IMPAIRMENT SCALE

☐ A = **Complete**: No motor or sensory function is preserved in the sacral segments S4-S5.

☐ B = **Incomplete**: Sensory but not motor function is preserved below the neurological level and includes the sacral segments S4-S5.

☐ C = **Incomplete**: Motor function is preserved below the neurological level, and more than half of key muscles below the neurological level have a muscle grade less than 3.

☐ D = **Incomplete**: Motor function is preserved below the neurological level, and at least half of key muscles below the neurological level have a muscle grade of 3 or more.

☐ E = **Normal**: Motor and sensory function are normal.

CLINICAL SYNDROMES (OPTIONAL)

☐ Central Cord
☐ Brown-Sequard
☐ Anterior Cord
☐ Conus Medullaris
☐ Cauda Equina

STEPS IN CLASSIFICATION

The following order is recommended in determining the classification of individuals with SCI.

1. Determine sensory levels for right and left sides.

2. Determine motor levels for right and left sides.
   Note: in regions where there is no myotome to test, the motor level is presumed to be the same as the sensory level.

3. Determine the single neurological level.
   This is the lowest segment where motor and sensory function is normal on both sides, and is the most cephalad of the sensory and motor levels determined in steps 1 and 2.

4. Determine whether the injury is Complete or Incomplete. (sacral sparing).
   If voluntary anal contraction = No AND all S4-S5 sensory scores = 0 AND any anal sensation = No, then injury is COMPLETE. Otherwise injury is incomplete.

5. Determine ASIA Impairment Scale (AIS) Grade:
   Is injury Complete?   If YES, AIS=A Record ZPP
   (for ZPP record lowest dermatome or myotome on each side with some (non zero score) preservation)

   NO → Is injury motor incomplete?   If NO, AIS=B
   (Yes=voluntary anal contraction OR motor function more than three levels below the motor level on a given side.)

   YES → Are at least half of the key muscles below the (single) neurological level graded 3 or better?

   NO → AIS=C   YES → AIS=D

   If sensation and motor function is normal in all segments, AIS=E
   Note: AIS E is used in follow up testing when an individual with a documented SCI has recovered normal function. If at initial testing no deficits are found, the individual is neurologically intact; the ASIA Impairment Scale does not apply.

Figure 7-2. International Standards for Neurological Classification of Spinal Cord Injury (ISCSCI) practice guidelines as published by The American Spinal Injury Association (ASIA).

PATHOANATOMY

The upper cervical spine anatomy consists of unique bone structures and intricate stabilizing structure (Figure 7-3). The occiput articulates with the atlas through paired condyles that form synovial joints with the lateral masses of the atlas. These paired joints are shallower and less well developed in children, thus contributing to the higher incidence of atlanto-occipital injuries in the pediatric population.[6] Other axis landmarks include the anterior and posterior tubercles. The anterior tubercle, located in the midline on the anterior arch, serves as the attachment site for the anterior longitudinal ligament and longus colli muscles. The posterior tubercle serves as the attachment site for the ligamentum nuchae. Lastly, the posterior arch of the atlas has a groove on the superior aspect that the vertebral artery courses through. The axis and atlas articulate through the odontoid process and posterior aspect of the atlas anterior arch, in addition to paired facet joints. These joints facilitate rotational range of motion.

The craniocervical ligamentous anatomy is typically divided into extrinsic (located outside of spinal canal) and intrinsic (located within spinal cord) ligaments. The extrinsic ligaments include the ligamentum nuchae, fibroelastic membranes, the atlanto-occipital joint capsule, and the atlanto-axial joint capsule. The intrinsic ligaments, provide the majority of the ligamentous stability. These ligaments form three layers: tectorial membrane, the cruciate ligament, and the odontoid ligaments. The tectorial membrane connects the posterior body of the axis to the anterior foramen magnum and is the cephalad continuation of the posterior longitudinal ligament. The cruciate ligament lies anterior to the tectorial membrane, behind the odontoid process. The transverse atlantal ligament is the strongest component, connecting the posterior odontoid to the anterior atlas arch, inserting laterally on bony tubercles. The odontoid ligaments include the alar and apical ligaments. The paired alar ligaments connect the odontoid to the occipital condyles, whereas the weaker apical ligament runs vertically between the odontoid and foramen magnum.

The major stabilizing structures between the occiput and upper cervical spine are the tectorial membrane and alar ligaments.[3] Hyperextension is limited by the tectorial membrane,

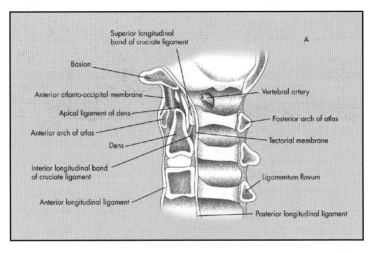

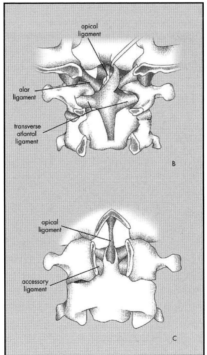

Figure 7-3. Upper cervical spine anatomy. (A) Occipital cervical articulation. (B) Posterior view of C1-2 articulation. (C) Anterior view of C1-2 articulation.

while rotation and latereal bending are restricted by the alar ligaments. Both the tectorial membrane and alar ligaments limit distraction. Hyperflexion is limited by bony anatomy.

IMAGING

A cervical spine plain radiograph series should be obtained in any patient where a c-spine injury is suspected. This series includes anteroposterior, lateral, and odontoid views. Good quality radiographs will include the superior portion of the T1 vertebral body. This quick, inexpensive study can provide reliable screening for blunt trauma patients.[7] The diagnostic performance of helical computed tomography (CT) scanners may be even better, with reported sensitivity as high as 99% and specificity 93%.[8,9] If the plain radiographs reveal any abnormality, or clinical suspicion is high, CT may be helpful in further defining the injury. CT also is helpful when the lower cervical spine cannot be visualized adequately on plain radiographs. Even if a CT is performed, plain radiographs should also be obtained to provide a baseline study for comparison when following these patients over time. There are some ligamentous injuries that may be missed by both plain radiographs and CT; these may only be detected with magnetic resonance imaging (MRI) or dynamic plain radiographs.

When interpreting imaging studies in patients with suspected cervical spine trauma, a systematic approach should be utilized. The AP view should be used to assess for alignment, visible fractures, and distraction. The odontoid view allows for assessment of the atlanto-axial complex. The dens should be evaluated for fracture on this view and the lateral masses of C1 should be aligned with the superior articular processes of C2 without significant overhang. The lateral view can be used to evaluate the anterior vertebral line, posterior vertebral line, and the spinolaminar line. All 3 lines should demonstrate a smooth curve with no step-off or discontinuity (Figure 7-4). Analysis of the prevertebral soft tissues on the lateral view should also be done; prevertebral soft tissue shadow should measure about 6 mm anterior to C2. Lastly, fractures at C1-C2 are associated with a remote subaxial cervical spine fracture, so imaging studies of these patients should be carefully reviewed.

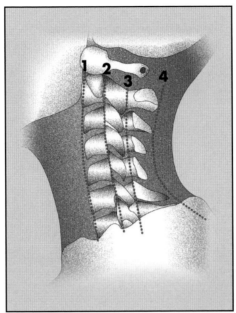

Figure 7-4. (1) Anterior vertebral, (2) posterior vertebral, (3) spinolaminal, and (4) interspinous lines.

Using lateral plain films, atlanto-occipital instability may be assessed using Power's ratio, Harris Rule of 12, Lee's X line, and the measurements of Wiesel and Rothman.[1,10,11] Power's ratio is commonly used to evaluate occipito-atlantal alignment. It is defined as the ratio of a line drawn from the basion to the anterior border of the posterior arch of the atlas and a line drawn from the opisthion to the posterior border of the anterior arch of the atlas (Figure 7-5). This ratio should be less than 1. A Power's ratio greater than 1 suggests an anterior atlanto-occipital dislocation. Additionally, the atlanto-dens interval (ADI) should be assessed to evaluate integrity of the atlanto-axial joint (see Figure 7-5). This measurement, made from the posterior aspect of the anterior C1 ring to the anterior cortex of the dens, should be less than 3 mm in adults and 4 mm in children. Higher values suggest disruption of the transverse ligament, so an MRI should be obtained to rule out pure soft tissue injury at this level.

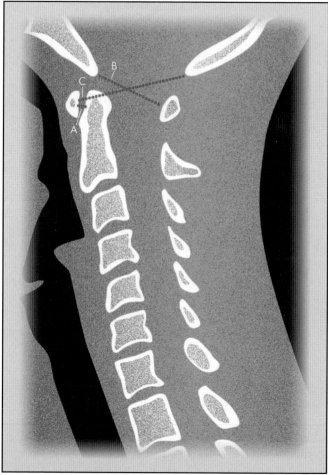

Figure 7-5. (A) ADI is calculated by drawing a line from the posterior aspect of the anterior arch of C1 to the most anterior aspect of the dens at the midpoint of the thickness of the arch. (B and C) Powers ratio is calculated as the ratio of a line drawn from the basion to the anterior border of the posterior arch of the atlas and a line drawn from the opisthion to the posterior border of the anterior arch of the atlas.

TREATMENT

Upper cervical spine trauma injuries are typically classified into 3 categories based on location: occipito-cervical, atlas, and axis. Occipital-cervical injuries include occipital condyle fractures and atlanto-occipital dislocation. Atlas injuries include various C1 fractures, while axis injuries include traumatic spondylolithesis and odontoid fractures.

Occipito-cervical

Occipital condyle fractures are usually associated with head trauma, skull base fractures, lower cranial nerve palsies (cranial nerve XII), and other upper cervical injuries.[12] These fractures can be very difficult to diagnose with X-ray alone and are better recognized on CT scan. Nondisplaced impaction and unilateral injuries can generally be treated with an orthosis for 6 to 12 weeks. Fracture patterns resulting in occipitocervical instability require halo or surgical stabilization.[13] The standard surgical intervention for this type of injury is a posterior occipitocervical instrumented fusion. Atlanto-occipital dislocation typically results from high energy trauma and accounts for 19% to 35% of all deaths from cervical spine trauma.[14] Children younger than 12 years of age are uniquely predisposed to this injury for 2 reasons: their atlanto-occipital joints are flatter and their head weight to body ratio is significantly greater than in adults.[15] Atlanto-occipital dislocations are virtually always unstable and should be expeditiously reduced and immobilized with a halo vest in patients too medically unstable to undergo immediate surgery. Traction should be avoided. Definitive treatment is a posterior instrumented occipito-cervical arthrodesis, which should be performed as soon as the patient is stable for surgery. Improved occipital plates and upper cervical poly-axial screw instrumentation techniques allow for excellent outcomes.[16,17]

Atlas (C1)

Five distinct fracture patterns are generally recognized (Figure 7-6). The most common injury is the posterior arch fracture (Gehweiler Type II), which is typically caused by hyperextension. Associated injuries of C2, such as an anterior

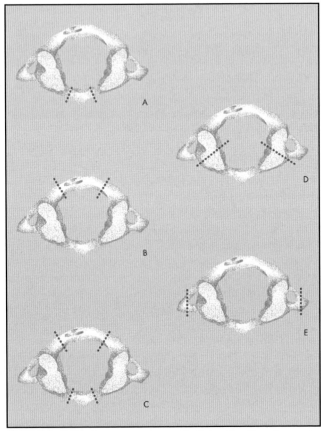

Figure 7-6. Atlas fracture patterns: (A) Gehweiler Type I: anterior arch, (B) Gehweiler Type II: posterior arch, (C) Gehweiler Type III: anterior and posterior arches (Jefferson fracture), (D) Gehweiler Type IV: lateral masses, and (E) Gehweiler Type V: transverse processes.

inferior body fracture or spondylisthesis of the axis, should be ruled out if this mechanism of injury is suspected. Jefferson fractures (Gehweiler Type III) (Figure 7-7) are characterized by fractures of the anterior and posterior arches that tend to displace laterally. Combined displacement of the lateral masses greater than 6.9 mm seen on the open mouth odontoid radiograph is associated with rupture of the transverse ligament. Treatment of most C1 injuries is nonoperative and

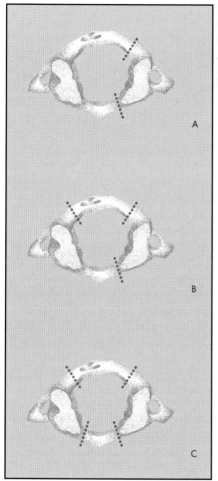

Figure 7-7. Jefferson fracture types: (A) 2-part, (B) 3-part, and (C) 4-part. The classic Jefferson fracture is a 4-part; however, the 2 other types may occur as well.

includes a cervical orthosis, Minerva cast,[11] or halo stabilization, typically utilized for 3 months. Contraindications to this treatment include patients with other spinal injuries and inability to comply with orthosis use. If the transverse ligament ruptures through its midportion, early C1-2 arthrodesis should be considered. Fractures of the posterior arch of the atlas by themselves are stable injuries and amenable to closed treatment with a cervical orthosis. Lateral mass fracture with minimal displacement may be treated with rigid orthosis or halo immobilization.

Axis (C2)

Traumatic spondylolisthesis of the axis, commonly called a Hangman's fracture, is characterized by a bilateral arch fracture of the C2 pars interarticularis with variable displacement of C2 on C3. The classification is as follows: Type I fractures have less than 3 mm of C2-3 subluxation; Type II fractures have more than 3 mm; Type IIA fractures have a larger C2-3 kyphotic angle compared to Type II; Type III fractures have the features of a Type II with bilateral facet dislocation. Type I, Type II and Type IIA fractures may usually be treated non-operatively in the acute setting, while Type III fractures should be addressed surgically. Ultimately, treatment of Hangman's fractures varies from institution to institution. Type I, stable Type II and IIA fractures can be treated by gentle reduction, sometimes necessitating several days of traction, followed by 12 weeks of collar or halo immobilization depending on the degree of instability. Successful treatment with hard collar immobilization alone has been reported. Unstable Type II and IIa may require surgical intervention.

The goals in surgical treatment of Type III Hangman's fractures are anatomic reduction and adequate stabilization. Surgical instrumentation for nonunion of a Type II Hangman's fracture has been described using both anterior and posterior approaches. The anterior approach, which has the advantage of technical ease and a relatively short fusion construct, involves a C2–C3 discectomy with interbody fusion and plating.[18-20] Several posterior options for fixation of the isthmus fracture component of a Hangman's fracture have also been described. Direct repair of the pars fracture with a screw across the fracture line has been reported to have the advantage of preserving motion of the axis.[21,22] However, direct pars repair does not address instability at the level of the disc, if present. Posterior C1–C3 wiring techniques have been described, but these techniques require postoperative halo immobilization and fusion involvement of C1, which is suboptimal. One of the best posterior fixation options in this setting is the use of a C2 pars or pedicle screw and a C3 lateral mass screw bilaterally. This technique addresses the detached posterior arch of C2 by pinning the fractured pars while simultaneously addressing instability at the C2, C3 disc.[23] Regardless of the choice of surgical approach, careful attention must be paid to achieving adequate reduction and stability. Outcomes in these patients appear to be affected by the quality of the reduction. Watanabe et al[24] reported that persistent post-operative neck pain may be related to residual kyphosis, residual translation and fracture extension into the C2 inferior facet.

Odontoid fractures constitute approximately 20% of cervical spine fractures.[25] Anderson and D'Alonzo created a classification system consisting of 3 types (Figure 7-8). Type I fractures involve the tip of the dens and may result from a severe rotational or lateral bending force that causes an avulsion of bone through the alar and apical ligaments. This type of fracture is stable and can be treated with a cervical orthosis in the absence of occipitocervical instability. Type III fractures extend into the body of C2 and have a much higher union rate than Type II fractures with nonoperative immobilization. This is likely due to a larger surface area and more vascular cancellous bone along the fracture line. This type of fracture may often be reliably managed with a halo fixator or cervical orthosis.[26,27]

Type II odontoid fractures occur through the "waist" of the dens without extension into the C2 body and have highly variable rates of nonunion. Risk factors for nonunion include age great than 50, more than 5 mm displacement, 9 degrees of angulation, fracture redisplacement, and smoking. Young patients with non-displaced or minimally displaced Type II that do not have without other significant risk factors for non-union, may be treated in a halo fixator. Orthosis treatment alone is reserved for elderly patients with low functional demand and should be undertaken with the expectation of a nonunion. Long term functional outcomes studies in elderly patients with nonunions are lacking in the literature.[23] Displaced and significantly angulated Type II fractures are often treated with posterior atlantoaxial instrumentation and fusion with autogenous bone graft or osteosynthesis with an anterior odontoid screw.[25] Posterior C1-2 fusion provides immediate, reliable stabilization and can be accomplished using wiring techniques, transarticular screw fixation, or segmental fixation utilizing C1 lateral mass and C2 pars or pedicle screws.[28] Posterior fusion techniques do not completely eliminate cervical rotation, so in select fractures, anterior fracture fixation should be

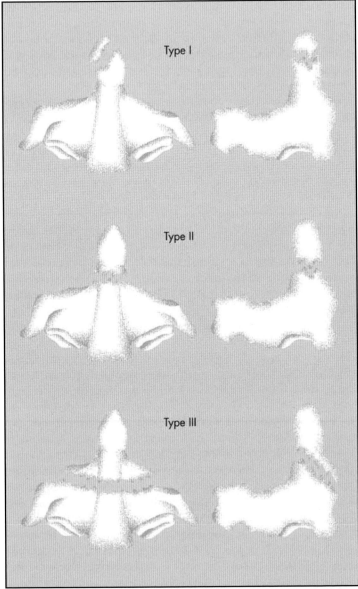

Figure 7-8. Odontoid fracture types. Three types of C2 odontoid fractures: Type I is an oblique fracture through the upper part of the odontoid process; Type II is a fracture occurring at the base of the odontoid as it attaches to the body of C2; Type III occurs when the fracture line extends through the body of the axis.

considered. Anterior screw fixation of acute odontoid fractures results in fusion rates over 90% and theoretically preserves C1-2 motion.[29] However, anterior screw fixation is technically demanding as it requires near-anatomic fracture reduction, absence of significant comminution, and proper orientation of the fracture plane. Midsubstance transverse ligament rupture, severe comminution, and an anterior-inferior to posterior-superior fracture pattern are contraindications to odontoid screw fixation.

C2 avulsion fractures involve a small portion of the anterior-inferior vertebral body are often referred to as extension teardrop fractures. These stable fractures should not be confused with subaxial flexion-compression teardrop fractures which are highly unstable injuries. C2 extension teardrop fractures can be treated with a hard collar.[30]

Conclusion

Upper cervical spine trauma is a significant cause of morbidity and mortality with diverse neurologic sequelae. A clear understanding of the mechanisms and patterns of cervical spine injury coupled with an appropriate exam and work-up allows timely and directed management of these injuries. Current technologies facilitate early diagnosis, and refined surgical techniques and instrumentation provide the best opportunity for returning trauma patients to pre-injury function (Table 7-2). Ligamentous injuries to the atlanto-occipital joint and transverse atlantal ligament are relatively uncommon, have a poor prognosis for healing, and typically respond best to surgical stabilization. Injury to bony structures, including occipital condyle fractures, atlas fractures, odontoid fractures, and traumatic spondylolisthesis of the axis, generally respond well to nonsurgical management.

Table 7-2

HELPFUL HINTS

Injury	Mechanism of Injury	Classification	Notes
Occipitocervical	Occipital condyle fracture	Axial compression (most common), high energy, blunt trauma	Type I fracture – Comminution from axial compression	CT scan to establish diagnosis and characterize fracture
			Type II fracture – Extension of a linear basilar skull fracture to include condyle	
			Type III – Alar ligament avulsion fracture	
	Occipitocervical instability	Multiple mechanisms	Type I injuries – Anterior occipital displacement	MRI to assess craniocervical ligaments. High morbidity and mortality.
			Type II injuries – Vertical displacement >2 mm between the occiput and C1	
			Type III injuries – Posterior occipital displacement	

(continued)

Table 7-2 (continued)

HELPFUL HINTS

Injury	Mechanism of Injury	Classification	Notes
Occipitocervical	Occipitocervical dissociation	Hyperextension with distraction		High mortality rate. Children are predisposed to these injuries.
Atlas (C1)	Fracture	Hyperextension (posterior arch), symmetric axial load (burst fracture), asymmetric (lateral mass fractures)	6 types: posterior arch, burst or Jefferson, lateral mass, anterior arch (blow-out or plow fractures), anterior tubercle, and transverse process.	High prevalence of concomitant upper c-spine injuires. Neurologic injury is rare.
	Transverse ligament rupture	Flexion force	Type I – Disruptions of the midsubstance Type II – Avulsions of the ligament from the C1 lateral mass	Maximum ADI is 3mm in adults and 5mm in adults. May occur in isolation or with other upper cervical injuries.
	Atlantoaxial rotary subluxation	Rotation and forward flexion		Dynamic CT for evaluation. Associated with impaction or avulsion injuries to the C1-2 articulation. *(continued)*

Table 7-2 (continued)

HELPFUL HINTS

Injury	Mechanism of Injury	Classification	Notes
Axis (C2)	Odontoid fracture	High velocity trauma and falls	Type I – Involving the tip of the dens	Type II fractures are the most common and have a higher non-union rate.
			Type II – Occurs through the waist without extension into the body	
			Type III – Entending into the C2 body	
	Traumatic spondylolithesis	Hyperextension with axial load, MVAs and falls	Type I – Fracture through neural arch; no angulation and minimal displacement.	Neurologic injury is uncommon due to fragment separation after fracture.
			Type IA – Elongation of the vertebral body; little angulation or translation.	
			Type II – Significant angulation and displacement.	
			Type IIA – Oblique fracture line; minimal translation but significant angulation.	
			Type III – Bilateral facet dislocation.	

(continued)

Table 7-2 (continued)

HELPFUL HINTS

Injury	Mechanism of Injury	Classification	Notes
Radiographs			
Plain X-ray			Screening radiographs should include AP, lateral, and open-mouth films. Supervised flexion and extension lateral radiographs may indicate instability at occiput-C1 and C1-2 articulations, however this can be dangerous if true instability exists.
CT			CT (thin slice) is very sensitive for evaluating the majority of suspected upper c-spine injuries, especially fractures.
MRI			MRI may be useful for patients with neural deficit, suspected transverse atlantal ligament injury, or other ligamentous injury.

REFERENCES

1. Ollerton JE, Parr MJ, Harrison K, Hanrahan B, Sugrue M. Potential cervical spine injury and difficult airway management for emergency intubation of trauma adults in the emergency department--a systematic review. *Emerg Med J.* 2006;23(1):3-11.
2. Patterson H. Emergency department intubation of trauma patients with undiagnosed cervical spine injury. *Emerg Med J.* 2004;21(3):302-305.
3. Jackson RS, Banit DM, Rhyne AL III, Darden BV II. Upper cervical spine injuries. *J Am Acad Orthop Surg.* 2002;10(4):271-280.
4. Ghanta MK, Smith LM, Polin RS, Marr AB, Spires WV. An analysis of Eastern Association for the Surgery of Trauma practice guidelines for cervical spine evaluation in a series of patients with multiple imaging techniques. *Am Surg.* 2002;68(6):563-567; discussion 567-568.
5. Mithani SK, St-Hilaire H, Brooke BS, Smith IM, Bluebond-Langner R, Rodriguez ED. Predictable patterns of intracranial and cervical spine injury in craniomaxillofacial trauma: Analysis of 4786 patients. *Plast Reconstr Surg.* 2009;123(4):1293-1301.
6. McCall T, Fassett D, Brockmeyer D. Cervical spine trauma in children: A review. *Neurosurg Focus.* 2006;20(2):E5.
7. Mower WR, Hoffman JR, Pollack CV Jr, Zucker MI, Browne BJ, Wolfson AB; NEXUS Group. Use of plain radiography to screen for cervical spine injuries. *Ann Emerg Med.* 2001;38(1):1-7.
8. McCulloch PT, France J, Jones DL, et al. Helical computed tomography alone compared with plain radiographs with adjunct computed tomography to evaluate the cervical spine after high-energy trauma. *J Bone Joint Surg Am.* 2005;87(11):2388-2394.
9. Holmes JF, Akkinepalli R. Computed tomography versus plain radiography to screen for cervical spine injury: A meta-analysis. *J Trauma.* 2005;58(5):902-905.
10. Powers B, Miller MD, Kramer RS, Martinez S, Gehweiler JA Jr. Traumatic anterior atlanto-occipital dislocation. *Neurosurgery.* 1979;4(1):12-17.
11. Wiesel SW, Rothman RH. Occipitoatlantal hypermobility. *Spine (Phila Pa 1976).* 1979;4(3):187-191.
12. Alcelik I, Manik KS, Sian PS, Khoshneviszadeh SE. Occipital condylar fractures. review of the literature and case report. *J Bone Joint Surg Br.* 2006;88(5):665-669.
13. Al-Khateeb H, Oussedik S. The management and treatment of cervical spine injuries. *Hosp Med.* 2005;66(7):389-395.
14. Hamai S, Harimaya K, Maeda T, Hosokawa A, Shida J, Iwamoto Y. Traumatic atlanto-occipital dislocation with atlantoaxial subluxation. *Spine (Phila Pa 1976).* 2006;31(13):E421-E424.
15. Bayar MA, Erdem Y, Ozturk K, Buharali Z. Isolated anterior arch fracture of the atlas: Child case report. *Spine (Phila Pa 1976).* 2002;27(2): E47-49.
16. Vaccaro AR, Lim MR, Lee JY. Indications for surgery and stabilization techniques of the occipito-cervical junction. *Injury.* 2005;(36 Suppl 2): B44-B53.

17. Reed CM, Campbell SE, Beall DP, Bui JS, Stefko RM. Atlanto-occipital dislocation with traumatic pseudomeningocele formation and post-traumatic syringomyelia. *Spine (Phila Pa 1976)*. 2005;30(5):E128-33.
18. Wang J, Jin D, Yao J, Chen J, Jiang J, Zhai D. Application of halo-vest in stable reconstruction of unstable upper cervical spine [in Chinese]. *Zhongguo Xiu Fu Chong Jian Wai Ke Za Zhi*. 2004;18(5):399-401.
19. Stulik J, Krbec M. Injuries of the atlas [in Czech]. *Acta Chir Orthop Traumatol Cech*. 2003;70(5):274-278.
20. Takahashi T, Tominaga T, Ezura M, Sato K, Yoshimoto T. Intraoperative angiography to prevent vertebral artery injury during reduction of a dislocated hangman fracture. case report. *J Neurosurg*. 2002;97(3 Suppl): 355-358.
21. Duggal N, Chamberlain RH, Perez-Garza LE, Espinoza-Larios A, Sonntag VK, Crawford NR. Hangman's fracture: A biomechanical comparison of stabilization techniques. *Spine (Phila Pa 1976)*. 2007;32(2): 182-187.
22. Bristol R, Henn JS, Dickman CA. Pars screw fixation of a hangman's fracture: Technical case report. *Neurosurgery*. 2005;56(1 Suppl):E204; discussion E204.
23. Platzer P, Thalhammer G, Ostermann R, Wieland T, Vécsei V, Gaebler C. Anterior screw fixation of odontoid fractures comparing younger and elderly patients. *Spine (Phila Pa 1976)*. 2007;32(16):1714-1720.
24. Watanabe M, Nomura T, Toh E, Sato M, Mochida J. Residual neck pain after traumatic spondylolisthesis of the axis. *J Spinal Disord Tech*. 2005;18(2):148-151.
25. Frangen TM, Zilkens C, Muhr G, Schinkel C. Odontoid fractures in the elderly: Dorsal C1/C2 fusion is superior to halo-vest immobilization. *J Trauma*. 2007;63(1):83-89.
26. Platzer P, Thalhammer G, Sarahrudi K, et al. Nonoperative manage-ment of odontoid fractures using a halothoracic vest. *Neurosurgery*. 2007;61(3):522-529; discussion 529-30.
27. Polin RS, Szabo T, Bogaev CA, et al. Nonoperative management of types II and III odontoid fractures: The philadelphia collar versus the halo vest. *Neurosurgery*. 1996;38(3):450-456; discussion 456-7.
28. Jeanneret B, Magerl F. Primary posterior fusion C1/2 in odontoid frac-tures: Indications, technique, and results of transarticular screw fixa-tion. *J Spinal Disord*. 1992;5(4):464-475.
29. Lee S, Chen J, Lee S. Management of acute odontoid fractures with single anterior screw fixation. *Journal of Clinical Neuroscience*. 2004;11:890-895.
30. Ianuzzi A, Zambrano I, Tataria J, et al. Biomechanical evaluation of sur-gical constructs for stabilization of cervical teardrop fractures. *Spine J*. 2006;6(5):514-523.

8

CERVICAL TRAUMA
SUBAXIAL CERVICAL
SPINE

Davor D. Saravanja, FRACS and Ian D. Farey, FRACS

INTRODUCTION

In North America, the incidence of adult cervical spine injuries after blunt trauma is 2% to 6%.[1] Subaxial cervical spine injuries can result in devastating outcomes for patients and their families and present a challenging problem for health care providers. These injuries can be associated with a spectrum of neurologic and orthopedic injury that may appear radiographically normal despite significant underlying instability. Missed injuries have the potential for iatrogenic deterioration and devastating complications and occur in up to 20% of cases.[2]

This chapter discusses a safe and systematic approach for examining the cervical spine, diagnosing subtle injuries,

Rihn JA, Harris EB. *Musculoskeletal Examination of the Spine: Making the Complex Simple* (pp. 125-148).
© 2011 SLACK Incorporated.

avoiding unnecessary studies, and providing appropriate care. Although much has been published on the clearance of cervical spine injury, the more cogent points of the literature are summarized, with an emphasis placed on higher-level evidence.

HISTORY

A high index of suspicion for a subaxial cervical injury should be present in all high-energy trauma victims. Patients who have associated injuries to the head or faciomaxillary regions have associated injuries to the cervical spine in up to 9% of cases.[3]

History particular to assessment of cervical trauma includes obtaining information on the mechanism of injury, the energy of the trauma, safety or restraining equipment used by the patient, and time of injury. Initial appearance of the patient at the scene of the trauma and initial and subsequent neurologic function can be crucial in decision making, particularly if there are no obvious injuries on initial examination. A large percentage of these patients (14%) are ambulatory at the time of presentation for medical treatment.[2] Significant existing medical conditions including ankylosing spondylitis, diffuse idiopathic skeletal hyperostosis (DISH), and connective tissue disorders as well as previous spinal surgeries also must be identified early to avoid missed injuries and identify those patients who may be more susceptible to spinal trauma.

EXAMINATION

Clinical examination of the patient with a suspected injury of the cervical spine begins with appropriate attention to safety (Table 8-1). In-line immobilization of the cervical spine in a rigid cervical collar and transport on a rigid spine board are now usual practice for paramedic transfers of trauma patients. In young children, a flat bolster beneath the shoulders is essential during transportation on a spinal board. Children have a disproportionately larger head than torso and hence have a risk of unnecessary stresses placed across the cervical spinal segments when immobilized without the bolster beneath their scapulae.

Table 8-1

METHODS FOR EXAMINING

Examination	Technique	Illustration	Grading	Significance
Safety	In-line immobilization in a rigid collar	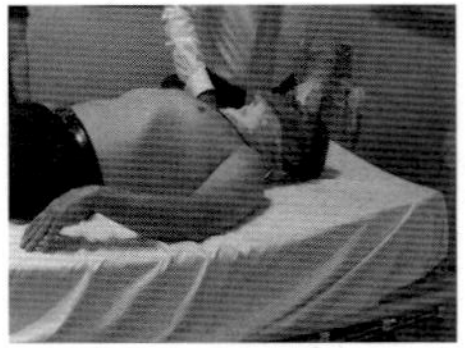		
Palpation	Supine with head controlled	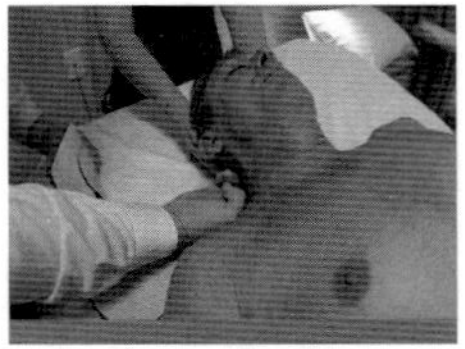		Midline tenderness is suggestive of a significant injury
	During log roll	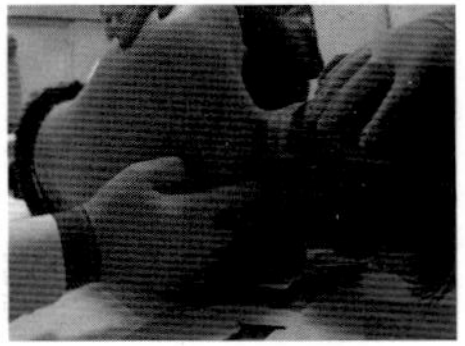		As above

(continued)

Table 8-1 (continued)

METHODS FOR EXAMINING

Examination	Technique	Illustration	Grading	Significance
Muscle grading	C5 biceps against resistance		Grade 0 – No contraction visible	Neural injury indicated by loss of strength
			Grade 1 – Visible contraction without movement	
			Grade 2 – Movement without gravity	
			Grade 3 – Movement through a full range against gravity	
			Grade 4 – Movement against resistance but not normal power	
			Grade 5 – Normal power	

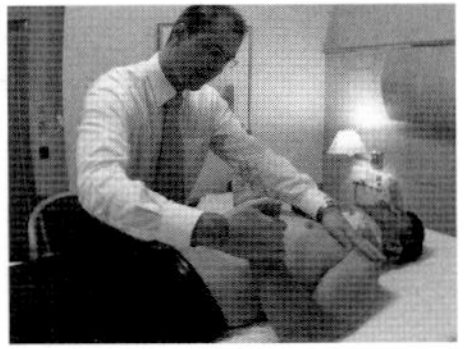

(continued)

Table 8-1 (continued)

Methods for Examining

Examination	Technique	Illustration	Grading	Significance
Muscle grading	C6 wrist extensors		See C5 biceps against resistance	See C5 biceps against resistance
	C7 triceps		See C5 biceps against resistance	See C5 biceps against resistance
	C8 finger flexors		See C5 biceps against resistance	See C5 biceps against resistance

(continued)

Table 8-1 (continued)

METHODS FOR EXAMINING

Examination	Technique	Illustration	Grading	Significance
Muscle grading	T1 finger abduction		See C5 biceps against resistance	See C5 biceps against resistance
Sensory level	C3 – T1		Grade 0 – Absent sensation Grade 1 – Altered sensation Grade 2 – Normal Sensation	Level of neurological injury can be determined
Reflexes	C5 biceps		0: Absent +: Present ++: Brisk	Objective sign of presence of muscle contraction. May be brisk in Myelopathy or upper motor neurone disorders

(continued)

Table 8-1 (continued)

METHODS FOR EXAMINING

Examination	Technique	Illustration	Grading	Significance
Reflexes	C6 brachioradialis		See C5 biceps	See C5 biceps
	C7 triceps		See C5 biceps	See C5 biceps

Rarely, patients may not have appropriate immobilization, or their cognitive state may prevent the maintenance of adequate immobilization when evaluated in the emergency department. It is essential to ensure adequate immobilization prior to commencing the examination. This may require intubation and general anesthesia in the emergency department for combative, intoxicated, or otherwise impaired patients.

Approximately 20% of patients who have a cervical spine fracture do not have midline cervical spine tenderness.[4] Heffernan et al[4] reported that all of the patients in their study with cervical fracture and no demonstrable midline cervical tenderness had an associated distracting upper torso injury. Upper torso injury was defined as an injury to the head, neck, face, upper extremity, chest, diaphragm, or thoracic spine. Lower torso injury involved the abdomen, pelvis, lower extremities, or lumbar spine. No patient with an isolated lower torso injury and the absence of midline cervical spine tenderness had a cervical spine fracture. When all of the patients with midline cervical spine tenderness were assessed in this series, 19.8% had a cervical fracture.[4]

The following list details the clinical examination of the cervical spine in the emergency department:

1. *Advanced Trauma Life Support (ATLS)* protocol assessment and treatment of airway (including cervical spine immobilization), breathing, and circulation

2. Inspection looking in particular for traumatic torticollis, previous surgical scars, open wounds, and associated head and neck trauma

3. Careful examination for midline tenderness[5] of the cervical spine by an emergency department physician or spine service physician (patient must be fully awake and alert–the presence of intoxication or sensory deficits precludes an accurate examination)

4. Examination of the paramedian muscles for tenderness and pain

5. A 3-person trauma log roll (neck immobilizing collar must be carefully removed) with attention to in-line support of the head, neck, and spine to visually inspect and palpate the entire length of the spine, followed by a digital rectal examination to assess distal anal sphincter function and tone, and the presence or absence of lower sacral sensation

Table 8-2

CERVICAL EXAMINATION BY RADICULAR BRANCHES

Spinal Level	Motor Examination	Sensory Examination	Reflex Test
C4		Lateral neck, suprascapular	
C5	Deltoid, biceps	Deltoid area	Biceps
C6	Wrist extensors	Radial dorsal hand, thumb	Brachioradialis
C7	Triceps	Middle finger	Triceps
C8	Finger flexors	Ulnar dorsal hand, little finger	
T1	Interossei (finger abduction)	Medial elbow and proximal forearm	

6. A careful neurologic examination of the upper and lower extremities (Table 8-2)
7. Exclusion of significant distracting injuries, predominantly in the upper torso

PATHOANATOMY

The subaxial cervical spine consists of the vertebrae from C3 through C7. In its uninjured natural state, it provides a lordotic curve protecting the cervical spinal cord and allowing passage of the spinal nerves from C3-C8 through its bilateral neural foramina. Significant motion is possible between these vertebrae. From C3 to C6, the vertebral arteries are contained in the osseous vertebral artery foramen of each transverse process.[6] Each of these vertebrae have a body, 2 transverse processes, 2 lateral masses that have sagittally inclined superior and inferior articular facets, laminae enclosing the posterior osseous arch of the spinal canal, and a spinous process that is typically bifid from C3-C6.

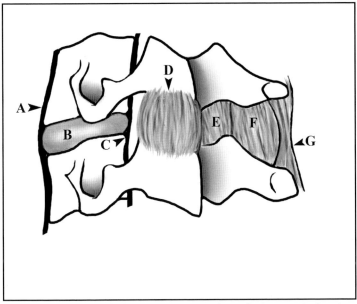

Figure 8-1. The diskoligamentous complex of the cervical spine. (A) Anterior longitudinal ligament, (B) intervertebral disk, (C) posterior longitudinal ligament, (D) facet joint capsule, (E) ligamentum flavum, (F) interspinous ligament, (G) supraspinous ligament.

The following is a list of the diskoligamentous complex (Figure 8-1), from anterior to posterior:

- Anterior longitudinal ligament
- Intervertebral disk
- Posterior longitudinal ligament
- Ligamentum flavum
- Facet joint capsule
- Interspinous ligament
- Supraspinous ligament

The ligamentum nuchae and cervical fascia are additional ligamentous structures that aid in the overall stability of the subaxial cervical spine. The posterior half of the diskoligamentous complex has been shown to be crucial in resisting kyphosis of the cervical spine in cadaveric sectioning models.[7] In cadaveric sectioning, the most robust of these structures was the capsule of the facet joints.

Pre-existing spinal fusion, whether congenital, acquired, or surgical, will lead to higher force transmission across residual mobile segments, predisposing to acute injury. Similarly, certain connective tissue disorders may predispose patients to traumatic subluxation and dislocation even in the absence of fractures. The consequences of missed cervical spine injury include kyphosis, pain, and neurologic deficit including the possibility of tetraplegia or quadriplegia.

IMAGING

The National Emergency X-Radiography Utilization Study studies recommend 5 guidelines that if present in isolation necessitate adequate imaging of the cervical spine: (1) midline cervical spine tenderness, (2) focal neurologic deficits, (3) altered level of consciousness, (4) evidence of intoxication,[8] or (5) a painful distracting injury.[9]

Asymptomatic patients are those patients who are neurologically normal with a Glasgow Coma Score of 15, no disorientation, no neck pain or distracting injuries, the ability to remember 3 objects at 5 minutes, no focal motor or sensory deficits, and appropriate responses to external stimuli. In this subset of patients, there is no need for radiographic clearance of the cervical spine.[10]

Radiographs

Cross-table plain radiographs miss up to 15% of blunt cervical vertebral injuries.[1] Oblique radiographs and flexion-extension imaging do not add any benefit to standard 3-view radiographs. The addition of a Swimmer's view or computed tomography (CT) of poorly visualized regions (eg, C7-T1) is necessary to complete adequate imaging. Three-view radiographs include a cross-table lateral aiming to include the C7-T1 articulation, an anteroposterior view of the cervical spine including the spinous process of C2, and an open-mouth odontoid view. The most common reasons for delayed or missed diagnosis include lack of experience in reading radiographs, inadequate radiographs, and incomplete radiographs,[2,6] typically those lacking full visualization of all vertebrae.

Radiographic signs of instability include the following:

- Loss of parallelism and/or uncovering of the cervical facet joints, and/or widened distance between spinous processes (Figures 8-2 to 8-5)
- Irregularity or step-off of the anterior spinal, spinolaminar, or posterior spinal lines (Figure 8-6)
- Minor fractures of the vertebral body (may represent reduced unilateral or bilateral facet dislocations) (see Figure 8-2)
- Soft-tissue swelling anterior to the cervical spine on a plain lateral radiograph >7 mm above C4 and >14 mm from C4-C7

Computed Tomography

CT is increasingly accessible and popular in imaging the cervical spine. Many authors propose that early CT of the whole cervical spine can supplant the use of plain radiographs.[11] Certainly, CT demonstrates excellent anatomic detail and can diagnose more occult fractures than plain radiographs; however, there is an increase in radiation exposure to the thyroid gland in particular.[12] Imaging of the craniocervical junction and the cervicothoracic junction is commonly performed by CT, especially when poorly visualized on plain radiographs or when there is a high index of clinical suspicion.

Magnetic Resonance Imaging

Magnetic resonance imaging (MRI) is recommended in the presence of cervical neurologic deficit for visualization of the neural elements as well as to assess the status of the posterior ligamentous complex. Some studies have found that MRI has the ability to diagnose injury to the diskoligamentous structures of the cervical spine; however, 2 recent studies have both demonstrated that the validity and positive predictive value of MRI is not adequate for diagnosis of such injury either anteriorly or posteriorly in the subaxial cervical spine.[13] MRI is helpful in distinguishing the degree of soft-tissue disruption in cervical facet subluxations and dislocations.

Figure 8-2. Case 1 Radiograph showing a C5-C6 unstable injury evidenced by loss of facet joint parallelism, anterior body corner fracture, loss of vertebral body height, and widened distance between C5 and C6 spinous processes.

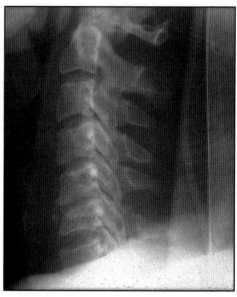

Figure 8-3. Case 2. Radiograph showing C6-C7 facet joint loss of parallelism indicating underlying instability. Note film is inadequate with C7-T1 not clearly visualized.

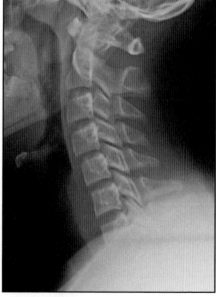

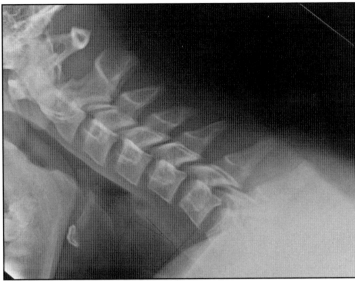

Figure 8-4. Case 2. Flexion radiograph of C6-C7 demonstrating subtle instability and uncovering of the facet joints at C6-C7.

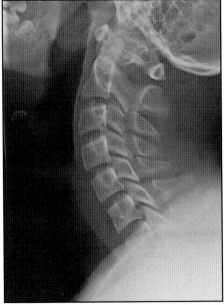

Figure 8-5. Case 2. Extension radiograph showing reduction of facet subluxation at C6-C7.

Figure 8-6. Case 1. Radiograph obtained 2 weeks following missed initial injury showing instability and subluxation at C5-C6.

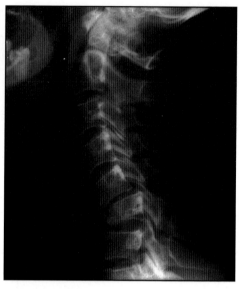

TREATMENT

The principles of treatment when managing subaxial cervical spine trauma are to prevent further injury, relieve compression causing neurologic compromise, restore anatomic stability and alignment, and preserve or restore function. There is a large spectrum of potential injury to the subaxial cervical spine, and definitive management guidelines are beyond the scope of this chapter. Pathology in the cervical spine can be treated operatively or nonoperatively. If operative management is recommended, then it can be performed through an anterior, posterior, or combined approach, depending on the requirements and goals of management.

In 2007, the Spine Trauma Study Group published an evidence-based algorithm for the treatment of subaxial cervical spine trauma.[14] The algorithm was based on their previously reported Subaxial Cervical Spine Injury Classification (SLIC) System (Table 8-3).[15] This section summarizes the principles as applied by the Spine Trauma Study Group.

Table 8-3

THE SUBAXIAL CERVICAL SPINE INJURY CLASSIFICATION (SLIC) SYSTEM

	Points
Morphology	
No abnormality	0
Compression + burst	$1 + 1 = 2$
Distraction (eg, facet perch or hyperextension)	3
Rotation or translation (eg, facet dislocation, unstable teardrop, or advanced stage flexion compression injury)	4
Diskoligamentous complex	
Intact	0
Indeterminate (eg, isolated interspinous widening, MRI signal change only)	1
Disrupted (eg, widening of anterior disk space, facet perch, or dislocation)	2
Neurologic status	
Intact	0
Root injury	1
Complete cord injury	2
Incomplete cord injury	3
Continuous cord compression (neuro modifier in the setting of a neurologic deficit)	+1

Reprinted with permission from Vaccaro AR, Hulbert RJ, Patel AA, et al. The subaxial cervical spine injury classification system: a novel approach to recognize the importance of morphology, neurology, and integrity of the disco-ligamentous complex. *Spine (Phila Pa 1976)*. 2007;32(21)2365-2374.

Central cord syndrome in the presence of cervical spondylosis is an injury pattern that is increasing in frequency as the population of older adults continues to increase. Surgery may be indicated in deteriorating incomplete neurologic pictures. The underlying cervical alignment and number of stenotic levels will dictate the surgical approach.[16] Focal kyphosis or anterior compression from acute intervertebral disk prolapse necessitates anterior decompressive surgery. This may include single or multiple anterior cervical diskectomy and fusion, or vertebrectomy and strut grafting. Neutral or lordotic posture of the cervical spine allows posterior decompression with either laminoplasty, or laminectomy and fusion. With the utilization of segmental instrumentation such as lateral mass screws, the ability to intraoperatively correct slight kyphosis into lordosis has improved.[14] Hence, a degree of discretion exists with choice of surgical approach that may improve outcomes, especially in the setting of ossification of the posterior longitudinal ligament.

Compression or burst fractures resulting from axial loading injuries may score highly if there is significant neurologic injury. Surgery may hence be the favored management option. In burst-type fractures, the compression is usually anterior as a result of the retropulsed fragments of the vertebral body. Anterior vertebrectomy and fusion directly decompresses the spinal cord and provides stability across the zone of injury. Subsequent posterior stabilization is rarely required in simple compression or burst injuries; however, in higher-energy burst fractures, it may be required due to the significant associated injury to the diskoligamentous complex.[17] Patients who demonstrate vertebral body retrolisthesis, facet dislocation, or translational shear injuries through the disk were more likely to require an anterior and posterior combined procedure.

Hyperextension injuries resulting in anterior diskoligamentous disruption typically require anterior surgical stabilization. Posterior augmentation over multiple levels is recommended when DISH (Figures 8-7 to 8-9) or ankylosing spondylitis are present due to the long lever arms of the multiple fused segments above and below the acute injury in these individuals.[14,18]

Unilateral or bilateral facet subluxations are complex injuries that may exhibit varying degrees of disruption and stability. In

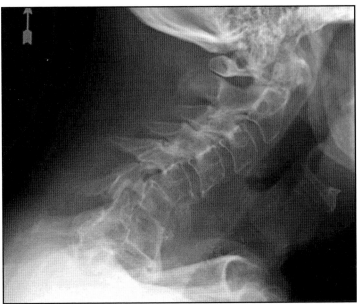

Figure 8-7. Case 3. DISH hyper-extension injury at 2 weeks post treatment in cervical orthosis, loss of anterior stability resulting in kyphotic deformity at C5-C6.

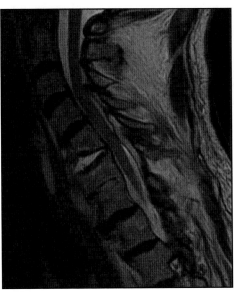

Figure 8-8. Case 3. MRI showing anterior injury to C5-C6 with associated canal stenosis and subtle increased cord signal associated with increased compression in kyphosis.

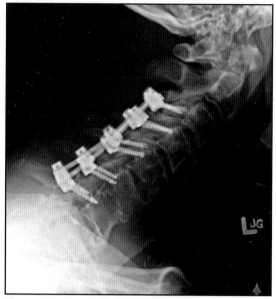

Figure 8-9. Case 3. Radiograph showing posterior multilevel fusion in presence of DISH and anterior C5-C6 injury.

the acute setting, the decision to apply traction and reduce dislocated injuries is still controversial. If there is extruded disk material behind the more caudal vertebral body, then closed reduction is not recommended,[19] and in neurologically compromised patients, emergent anterior decompressive surgery should be performed to remove the disk fragment prior to reduction. Stand-alone anterior fixation can be used if a congruous reduction of the facet joints is achieved.[20] In the absence of disk material behind the vertebral body, one can choose to operate either anteriorly or posteriorly depending on surgeon preference and injury morphology.[14] When performing a posterior stabilization alone, any torn ligamentum flavum or hematoma should be debrided prior to applying compression across the construct. This will decrease the risk of cord compression from infolding of these soft tissues.[21]

Unilateral or bilateral facet fracture dislocation or subluxation results in rotational or translational injuries. These are always associated with a degree of diskoligamentous injury and hence score highly on the SLIC classification prior to factoring neurology.[14] If there is an associated end-plate vertebral body fracture, then the use of anterior only stabilization is contraindicated.[22] However, if there is no vertebral body fracture, then either anterior or posterior surgery may be performed depending on the specific fracture morphology. Anterior diskectomy and fusion are recommended when there is evidence of disk material in the canal. In the absence of loose disk material in the canal, either anterior or posterior stabilization can be performed. Combined anterior and posterior surgery is recommended if there is concern for the development of kyphosis based on posterior diskoligamentous complex injury.[23]

CONCLUSION

Cervical spine trauma below the axis is common in blunt trauma. The need for rapid, safe, and accurate assessment is paramount as missed injuries can lead to catastrophic loss of function. Although there are many forms of subaxial cervical injury, the use of standardized terminology in the form of the SLIC system enables a better understanding of the various injuries, allowing efficacious and safe management by spine surgeons (Table 8-4).

REFERENCES

1. Griffen MM, Frykberg ER, Kerwin AJ, et al. Radiographic clearance of blunt cervical spine injury: plain radiograph or computed tomography scan? *J Trauma*. 2003;55(2):222-227.
2. Platzer P, Hauswirth N, Jaindl M, et al. Delayed or missed diagnosis of cervical spine injuries. *J Trauma*. 2006; 61(1):150-155.
3. Elahi MM, Brar MS, Ahmed N, et al. Cervical spine injury in association with craniomaxillofacial fractures. *Plast Reconstruct Surg*. 2008;121(1): 201-209.
4. Heffernan DS, SchermerCR, Lu SW. What defines a distracting injury in cervical spine assessment? *J Trauma*. 2005;59(6):1396-1399.
5. Hoffman JR, Mower WR, Wolfson AB, Todd KH, Zucker MI. Validity of a set of clinical criteria to rule out injury to the cervical spine in patients with blunt trauma. National Emergency X-Radiography Utilization Study Group. *N Engl J Med*. 2000;343(2):94-99.

Table 8-4

HELPFUL HINTS

Examination Technique	Findings	Actions
1. Attention to safety	Alert and cooperative patient	ATLS assessment and Cervical orthosis applied with in-line immobilization
	Impaired cognitive status but noncombative and not restless	As Above Observe closely for change in status Imaging required
	Impaired cognitive status and restless or combative	Assess for need of urgent intubation and general anaesthesia to aid protection of cervical spine Imaging required
2. Inspection	Penetrating neck trauma, Torticollis or spasms, Head or neck trauma, or scars suggesting previous surgery	Imaging required
3. Midline cervical spine palpation with in-line immobilization	No midline tenderness; no upper torso distracting injury	Clinical clearance is possible if all NEXUS criteria satisfied
	Tenderness in midline Patient awake and alert and not impaired	Imaging required
4. Neurologic assessment	No deficits	Assess all other findings prior to clearance of injury
	Neurological deficit with or without sensory levels	Indicates level of neurological injury, imaging required

(continued)

Table 8-4 (continued)

HELPFUL HINTS

Examination Technique	Findings	Actions
5. 3-person log roll	No visual sign of injury, no tenderness in midline or paraspinal muscles, normal neurological and rectal examination	May be suitable for clinical clearance as per NEXUS criteria
	Visual signs of trauma such as bruising or penetrating injury, tenderness, or neurological deficit	Imaging required

IMAGING

Modality	Appropriate Views	Findings
X-ray	3 view trauma series:	Obvious fractures
	Cross table lateral (must include C7-T1 joints),	Soft tissue swelling anterior to the cervical spine
	AP and, open-mouth odontoid projections.	Facet joint irregularity
		Facet dislocations
	Swimmer's views may help visualize the lower cervical spine	
CT	Axial scans, 2 dimensional sagittal and coronal reconstruction, may have 3 dimensional reconstructions	Highly detailed depiction of fractures and facet alignment including diagnosis of subtle fractures
		Cervico-thoracic junction easily visualized
MRI	Sagittal and axial images, T2 and STIR sequences often most helpful	Assessment of displaced injuries requiring reduction to visualize disc material behind vertebrae and guide operative intervention

6. Kwon BK, Vaccaro AR, Grauer JN, Fisher CG, Dvorak MF. Subaxial cervical spine trauma. *J Am Acad Orthop Surg.* 2006;14(2):78-89.
7. White AA, Panjabi MM. *Clinical Biomechanics of the Spine.* Philadelphia, PA: Lippincott; 1990.
8. Harris MB, Kronlage SC, Carboni PA, et al. Evaluation of the cervical spine in the polytrauma patient. *Spine (Phila Pa 1976).* 2000;25(22): 2884-2892.
9. Hoffman JR, Wolfson AB, Todd KH, Mower WR. Selective cervical spine radiography in blunt trauma: methodology of the National Emergency X-Radiography Utilization Study (NEXUS). *Ann Emerg Med.* 1998;32(4):461-469.
10. Radiographic assessment of the cervical spine in asymptomatic trauma patients. *Neurosurgery.* 2002; 50(3)(suppl):S30-S35.
11. Mathen R, Inaba K, Munera F, et al. Prospective evaluation of multislice computed tomography versus plain radiographic cervical spine clearance in trauma patients. *J Trauma.* 2007; 62(6):1427-1431.
12. Banit DM, Grau G, Fisher JR. Evaluation of the acute cervical spine: a management algorithm. *J Trauma.* 2000;49(3);450-456.
13. Vaccaro AR, Rihn J, Fisher CG, et al. A prospective analysis comparing pre-operative MRI to diagnose disruption of the posterior ligamentous complex in cervical spine injury and intraoperative findings. Paper presented at: Spine Society of Australia Annual Scientific Meeting; April 18, 2009; Brisbane, Queensland, Australia.
14. Dvorak MF, Fisher CG, Fehlings MG, et al. The surgical approach to subaxial cervical spine injuries: an evidence-based algorithm based on the SLIC classification system. *Spine (Phila Pa 1976).* 2007;32(23): 2620-2629.
15. Vaccaro AR, Hulbert RJ, Patel AA, et al. The subaxial cervical spine injury classification system: a novel approach to recognize the importance of morphology, neurology, and integrity of the disco-ligamentous complex. *Spine (Phila Pa 1976).* 2007;32(21):2365-2374.
16. Harrop JS, Sharan A, Ratliff J. Central cord injury: pathophysiology, management, and outcomes. *Spine J.* 2006;6(6 Suppl):S198-S206.
17. Cybulski GR, Douglas RA, Meyer PR Jr, et al. Complications in three-column cervical spine injuries requiring anterior-posterior stabilization. *Spine (Phila Pa 1976).* 1992;17(3):253-256.
18. Einsiedel T, Schmelz A, Arand M, et al. Injuries of the cervical spine in patients with ankylosing spondylitis: experience at two trauma centers. *J Neurosurg Spine.* 2006;5(1):33-45.
19. Vacarro AR, Falatyn SP, Flanders AE, et al. Magnetic resonance evaluation of the intervertebral disc, spinal ligaments, and spinal cord before and after closed traction of cervical spine dislocations. *Spine (Phila Pa 1976).* 1999;24(12):1210-1217.
20. Aebi M, Zuber K, Marchesi D. Treatment of cervical spine injuries with anterior plating. Indications, techniques, and results. *Spine (Phila Pa 1976).* 1991;16(3 Suppl):S72-S79.
21. Ludwig SC, Vaccaro AR, Balderston RA, et al. Immediate quadriparesis after manipulation for bilateral cervical facet subluxation: a case report. *J Bone Joint Surg Am.* 1997;79(4):587-590.

22. Johnson MG, Fisher CG, Boyd M, et al. The radiographic failure of single segment anterior cervical plate fixation in traumatic cervical flexion distraction injuries. *Spine (Phila Pa 1976)*. 2004;29(24):2815-2820.

23. Payer M. Immediate open anterior reduction and antero-posterior fixation/fusion for bilateral cervical locked facets. *Acta Neurochir (Wein)*. 2005;147(5):509-513.

9

LUMBAR DISK
HERNIATION

Mark L. Dumonski, MD and D. Greg Anderson, MD

INTRODUCTION

As lumbar intervertebral disks age, multiple biomechanical changes occur that result in alterations in their structural properties. This inevitable process, termed disk degeneration, has various implications. As the degenerative process ensues, a weakening of the supporting structures occurs, including a loss of cohesion of the molecules in the nucleus pulposus and fissuring of the annulus fibrosus. This sets the stage for herniation of the nucleus pulposus through the annulus fibrosus and into the spinal canal or adjacent to the lumbar nerve roots. Lumbar disk herniation is one of the most common diagnoses encountered

Rihn JA, Harris EB. *Musculoskeletal Examination of the Spine: Making the Complex Simple* (pp. 149-166).
© 2011 SLACK Incorporated.

in a typical spine clinic. The exact prevalence of the condition has been difficult to estimate, as some degree of disk herniation may be seen in 10% to 81% of asymptomatic patients.[1] However, the lifetime prevalence of symptomatic disk herniation leading to radiculopathy is approximately 1%. Although any level in the spine can be affected, the L4-L5 and L5-S1 levels are responsible for approximately 85% to 90% of all lumbar herniations.

Various monozygotic twin studies have been performed in an attempt to elucidate risk factors associated with accelerated rates of degeneration. Although there are undoubtedly various risk factors yet to be diskovered, a few are known to play a role in accelerating the process of disk degeneration and ultimately disk herniation. Among the known risk factors are a history of heavy lifting, smoking, and age. However, the strongest influence in the advancement of disk degeneration is genetics.[2,3] Genetics account for 61% of the variance at the T12-L4 levels and for 41% from L4-S1.

HISTORY

Patients with symptomatic disk herniations may ultimately seek medical attention as a result of either back pain or leg pain, called radiculopathy or sciatica. Patients with an acute herniation can often recall a specific event that resulted in pain (eg, falling on ice or lifting a box). The sitting position generates the highest intradiskal pressures compared to other positions[4] and thus pain is typically most prominent in the sitting position, although patients with severe sciatica often complain of severe leg pain while walking.

Although back pain may be vague, nonspecific, and difficult to treat, radicular pain is more specific and more readily treatable. Patients often complain of pain in a specific dermatomal distribution (Figure 9-1). For example, a patient with a disk herniation compromising the S1 nerve root would typically complain of pain along the posterior calf and lateral foot, whereas a patient with L5 radiculopathy will commonly have pain in the lateral leg and dorsum of the foot. It is also important to ask patients about any weakness that they might be experiencing. Patients with L4 or L5 radiculopathy may have difficulty in getting the foot to clear the floor during the swing phase of gait,

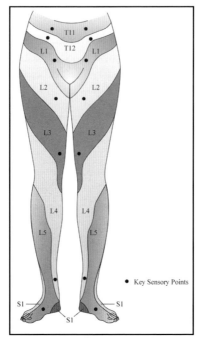

Figure 9-1. Illustration of lower-extremity dermatomes.

whereas patients with S1 radiculopathy might notice weakness during step-off. It is important to note that although radicular pain is a common complaint related to a disk herniation, the dermatomal distribution of a patient's pain may not always coincide with any particular "classic" dermatome.

PHYSICAL EXAMINATION

The patient's gait should first be evaluated. As flexion of the spine has been found to increase the size of the foramen,[5] patients with foraminal nerve root compression typically learn to assume a forward-flexed posture to minimize symptoms. Similarly, patients may also be noted to lean away from the side with a compressed nerve root, as this also aids in increasing the foraminal space available for the exiting nerve. Patients

L2			Hip flexors
L3			Knee extensors
L4			Ankle dorsiflexors
L5			Long toe extensors
S1			Ankle plantar flexors

Figure 9-2. Summary of key lower-extremity muscle groups.

should be instructed to walk on their heels, then on their toes. This is a sensitive means to identify subtle weakness in the ankle dorsiflexors or plantarflexors that might otherwise be missed. Also, a Trendelenburg gait may be noted, suggesting an L5 palsy (affecting the hip abductors).

After assessing gait, the spine should then be palpated both at the midline and in the area of the paraspinal musculature. Paraspinal tenderness is suggestive of muscle spasm, whereas midline tenderness is generally nonspecific.

Reflexes should be evaluated and any asymmetry should be noted. Increased tone is suggestive of a pathologic condition of the central nervous system, such as spinal cord compression in the cervical or thoracic spine or intracranial pathology; decreased tone is suggestive of nerve root compression. Decreased or increased reflexes that are symmetric may be normal. Manual muscle testing should then be evaluated thoroughly, again noting any asymmetry (Figure 9-2). It is important to hold counter pressure for 2 to 3 seconds for each key muscle group that is tested, as this is a more sensitive means of detecting subtle weakness. Sensation is then tested by lightly stroking each dermatome bilaterally.

The patient is then positioned supine and a straight-leg raise is performed by passively elevating the leg 60 degrees (Table 9-1; Figure 9-3). This test stretches the nerve that is tented over the herniated disk and exacerbates the radicular symptoms. This test is approximately 67% sensitive in diagnosing pathology involving the L4, L5, or S1 nerve roots,[6] as these are the nerves placed on tension during this maneuver. It is considered positive if the patient's radicular pain is reproduced;

Table 9-1

METHODS FOR EXAMINATION

Examination	Technique	Grading	Significance
Straight-leg raise	The patient is placed supine on an examination table and the thigh is passively flexed with the knee extended	Positive when radicular symptoms are reproduced past the knee	Suggestive of a lower lumbar disk herniation (L3/4, L4/5, L5/S1)
Femoral nerve stretch test	The patient is placed in the lateral decubitus position and the thigh is passively extended with the knee flexed 90 degrees	Positive when radicular symptoms are reproduced into the anterior thigh	Suggestive of an upper lumbar disk herniation (L1/2, L2/3)

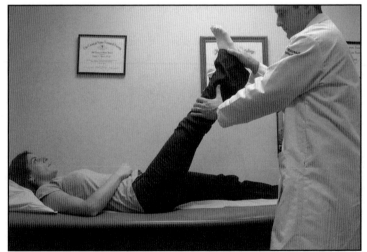

Figure 9-3. A straight-leg raise is performed to evaluate radicular pain.

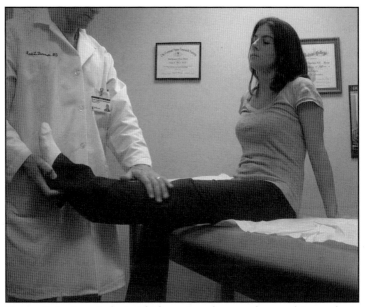

Figure 9-4. A seated straight-leg raise also can be used to evaluate radicular pain.

reproduction of back pain is not a positive test. Alternatively, a seated straight-leg raise may be performed (Figure 9-4), in which the knee is extended while the patient is in the seated position. Although a seated straight-leg raise is easier to perform, it is much less sensitive (41%) than a straight-leg raise with the patient supine. Elevating the contralateral leg may also produce symptoms on the ipsilateral side. This generally suggests an axillary disk herniation.

The femoral nerve stretch test is then performed to test for a disk herniation affecting the upper lumbar nerve roots (ie, L2 and L3). This is performed by placing the patient on the contralateral side, cradling the ipsilateral leg with the knee flexed 90 degrees, and passively extending the hip (Figure 9-5).

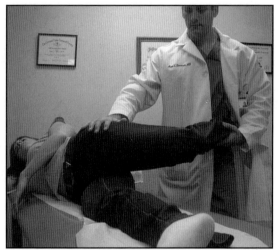

Figure 9-5. The femoral nerve stretch test is performed to test for a disk herniation affecting the upper lumbar nerve roots.

PATHOANATOMY

The intervertebral disk forms the anterior boundary of the spinal canal at the level of the facet joints and intervertebral foramina. Overlying the posterior vertebral bodies and posterior disk margins is the posterior longitudinal ligament. The posterior longitudinal ligament is a relatively thin structure. It tapers additionally as it expands laterally across the intervertebral disk and is less developed at the more caudal lumbar levels.[7] These observations are thought to contribute to the higher rate of herniations in the posterolateral region at the lower lumbar levels (ie, L4-L5 and L5-S1; Figure 9-6).

An understanding of the anatomic relationships within the lumbar spine is integral in understanding the neurologic consequences of lumbar disk herniations and the operative approaches required to address them. Nerve roots typically depart from the cauda equina one level above the foramen in which they exit. For example, the L4 nerve root typically departs the cauda equina at the level of the L3 vertebral body. It then courses inferiorly across the L3-L4 intervertebral disk (anteriorly) and the L3-L4 facet (posteriorly), beneath the L4

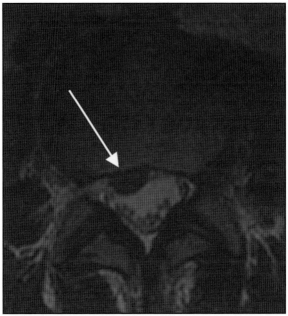

Figure 9-6. Axial T2-weighted MRI demonstrating a typical posterolateral disk herniation (arrow).

pedicle, and finally exits the L4-L5 foramen (Figure 9-7). The typical posterolateral disk herniation affects the root of the lower numbered segment. In contrast, a foraminal or far lateral disk herniation affects the nerve exiting the foramen above the disk space, which is the nerve of the upper numbered segment.

The coronal plane can be divided into 3 distinct zones (see Figure 9-7). The central zone is located between the 2 lateral margins of the cauda equina. Here, the intervertebral disk is found immediately anterior and the ligamentum flavum is posterior. The lateral recess is between the lateral margin of the cauda equina and the medial border of the pedicle. Here, the intervertebral disk is anterior and the superior articular process of the vertebral segment below is immediately posterior. Finally, the foraminal zone is between the medial and lateral borders of the pedicle. Herniations may also be found lateral to the pedicle. These are termed either far-lateral or extraforaminal herniations. Herniations into the central zone or lateral recess are expected to affect the traversing nerve root

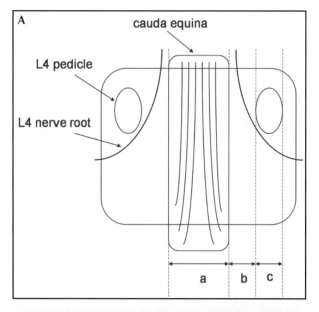

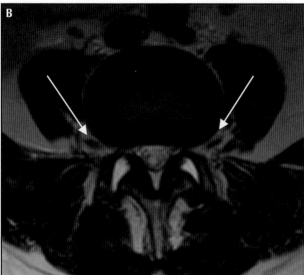

Figure 9-7. (A) Schematic demonstrating the anatomic relationships between the exiting nerve root, the pedicle, and the cauda equina, as well as the 3 anatomic zones: a–central zone, b–lateral recess, and c–foraminal zone. (B) Axial T2-weighted MRI demonstrating the L4-L5 disk and the exiting L4 nerve roots (arrows). A typical posterolateral herniation at this level would not affect the exiting nerve root, as it is shown to have already exited the spinal canal and foramen.

(ie, the L5 nerve root in an L4-L5 herniation), whereas foraminal and extraforaminal herniations will commonly affect the exiting nerve root (Figure 9-8).

IMAGING

Radiographs

Plain radiographs are routinely the first imaging modality to be evaluated. Although plain radiographs are unable to detect a disk herniation, there are various degenerative osseous findings commonly associated with a disk herniation that may be present. These include narrowing of the disk space, osteophytes, hypertrophy of the facets, spondylolisthesis, and lateral listhesis. Flexion-extension views may aid in detecting subtle segmental instability not otherwise noted on static films. Although these findings may be helpful in accounting for degenerative changes, they do not diagnose a disk herniation.

Magnetic Resonance Imaging

Magnetic resonance imaging (MRI) is the study of choice in detecting lumbar disk herniations, with a reported sensitivity and specificity of 92% and 100%, respectively.[8] Sagittal and axial T2-weighted images are typically most helpful in characterizing a herniation, whereas parasagittal T1- and T2-weighted images can be useful in evaluating the status of the foramina (see Figure 9-8B).

Morphologically, disk herniations are described as either protruded, extruded, or sequestered.[9] A protrusion is an eccentric bulge within an intact annulus. A protrusion involving <25% of the disk circumference is focal, whereas a broad-based protrusion involves 25% to 50% of the circumference (Figure 9-9). An extrusion is present when the diameter of the herniated fragment beyond the disk space is greater than the diameter of the fragment at its base. A sequestered fragment is one in which there is no continuity between the fragment and the disk of origin (ie, a free fragment). Finally, a contained herniation is located within an intact outer annulus, whereas an uncontained fragment breaches the annulus.

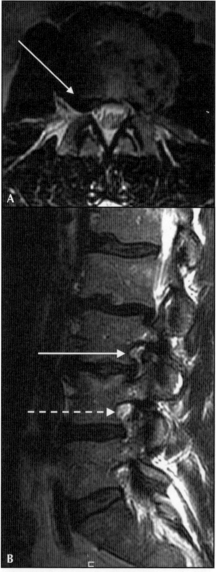

Figure 9-8. (A) Axial and (B) parasagittal T2-weighted MRIs demonstrating a foraminal L3-L4 disk herniation. On the parasagittal view, note the stenotic L3-L4 foramen (solid white arrow) compared to the patent L4-L5 foramen below (dashed white arrow).

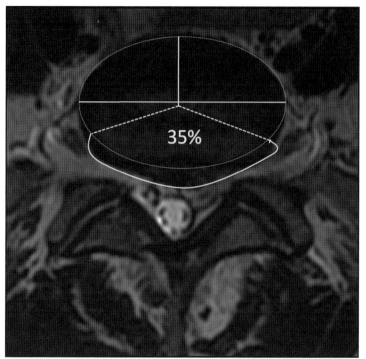

Figure 9-9. Axial T2-weighted MRI demonstrating a broad-based disk protrusion measuring 35% of the circumference of the disk.

Computed Tomography Myelography

Computed tomography myelography (CTM) is the test of choice in patients who are unable to undergo MRI. The most common reasons patients are unable to undergo MRI include severe claustrophobia, the presence of a pacemaker, or morbid obesity. With CTM, dye is injected into the thecal sac and a computed tomography is obtained. Although the soft tissues are poorly visualized, extradural compression of the thecal sac due to a soft-tissue mass is consistent with a disk herniation. Due to the inferiority of evaluating the soft tissues directly, the sensitivity and specificity in diagnosing a disk herniation using CTM are slightly inferior (83% and 71%, respectively) to MRI.[8]

TREATMENT

Nonoperative Treatment

Patients presenting with an acute disk herniation may have significant back and leg pain. In this acute setting, bed rest may be advised but should be limited to no more than 2 to 3 days. Narcotics are also appropriate acutely but should not be continued for more than a few days. After this time frame, patients should be encouraged to pursue light activity as tolerated. It is important to reassure patients that although they may be experiencing significant diskomfort, they are unlikely to cause additional damage or dysfunction. In addition, prolonged immobilization is not without consequence and can have a number of deleterious effects (eg, weight gain, depression, altered pain thresholds, and deep venous thrombosis).

Additional nonoperative treatment modalities should be aimed at restoring the patient's function back to his or her preinjury level. Options include nonsteroidal anti-inflammatory drugs, oral or injected steroids, and physical therapy. For patients who do not respond to rest and nonsteroidal anti-inflammatory drugs, injected steroids in the form of a transforaminal epidural or selective nerve root block can be helpful in reducing radicular pain and aiding in confirmation of the diagnosis. Patients responding favorably to an injection at a particular level are more likely to respond favorably to surgical intervention should it be required in the future. Patients with radicular pain who receive selective nerve root injections have an 85% chance of having a successful outcome by the 1-year time point, whereas 10% to 15% will ultimately require surgery.[10-13]

As mentioned previously, the majority of patients with radiculopathy will not require surgery. Thus, conservative treatment should always be the first line of treatment unless the patient is experiencing severe neurologic symptoms as a result of the disk herniation. In rare cases, a large disk herniation may cause severe compression of the thecal sac, leading to progressive dysfunction of the entire cauda equina, leading to cauda equina syndrome (Figure 9-10).

Signs and symptoms of cauda equina syndrome include progressive bilateral lower extremity and perineal weakness

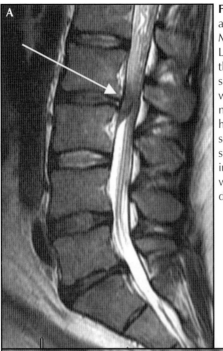

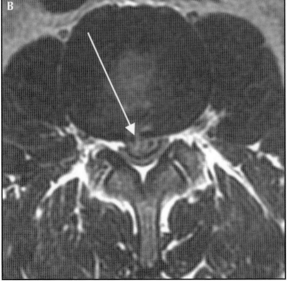

Figure 9-10. (A) Sagittal and (B) axial T2-weighted MRIs demonstrating a large L2-L3 disk herniation. On the axial view, the thecal sac is severely compressed with no cerebrospinal fluid noted at the level of the herniation. This patient presented with cauda equina syndrome 4 hours after lifting a heavy box and underwent emergent surgery for decompression.

(decreased rectal tone), lower extremity and perineal sensory loss (saddle anesthesia) and decreased reflexes, in addition to bowel and bladder incontinence. Cauda equina syndrome is considered a true surgical emergency. Although most patients can be expected to regain function following decompression (rates for the return of continence, sexual function, and rectal function are 73%, 67%, and 64%, respectively), it has been shown that patients undergoing decompression within 48 hours are more likely to regain function compared to patients who undergo decompression after 48 hours.[14] Although it has not been determined conclusively, it stands to reason that the earlier the decompression, the more favorable the patient's functional recovery.[15]

Operative Treatment

Patients failing conservative treatment for at least 6 weeks are candidates for operative intervention. If surgery is considered, it is important for patients to understand what they can expect postoperatively and how this compares to nonoperative treatment. Although surgery can be expected to provide a significant improvement in symptoms early in the postoperative period, the longer-term results are not significantly different between operative and nonoperative candidates.[16-18]

Various surgical diskectomy techniques are described to achieve fragment removal and decompression of the neural elements. In general, a midline 1- to 2-inch incision is made over the indicated segment, followed by a fascial incision 1- to 2-mm lateral to the midline on the side of the herniation (to allow reapproximation of the fascia). A laminotomy is then performed, removing bone from the cephalad and caudad hemilamina.

After the ligamentum flavum is identified, it is also removed, allowing visualization of the dura (medially) and the traversing nerve root (laterally). The neural elements are then gently retracted medially, gaining access to the disk space and herniated fragment (Figure 9-11). If the fragment is contained, a small annulotomy is made and the fragment is removed.

Alternatively, a Wiltse approach (1- to 2-inches lateral of the midline) may be used to access far lateral herniations. This approach is carried between multifidus (medially) and longissimus (laterally), through the intertransverse membrane,

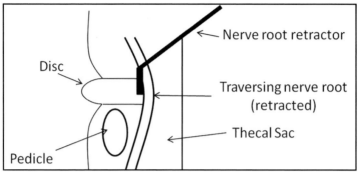

Figure 9-11. Illustration showing the intraoperative view during a routine diskectomy. Note the medial retraction of the traversing nerve root and thecal sac that is routinely required for adequate visualization of the herniated fragment.

and finally to the herniated fragment. Patients are generally released home following the procedure and can resume activities as tolerated without restriction.

Conclusion

Lumbar disk herniations are common (Table 9-2). When symptomatic, significant pain and disability may result. The patient's history, physical examination, and imaging are typically diagnostic. Posterolateral lumbar disk herniations typically affect the nerve root traversing the herniated fragment (the nerve root of the lower numbered segment). Selective nerve root blocks can play both a diagnostic and therapeutic role for patients with disk herniations. With the exception of cauda equina syndrome, which should be treated emergently, radiculopathy secondary to a lumbar disk herniation should be treated with nonoperative modalities for at least 6 weeks. If a patient fails nonoperative management, surgery is generally helpful in relieving the radicular symptoms. Patients should be educated about the natural history of disk herniations, including the fact that nonoperative treatment generally yields equivalent results in the long term compared to surgical treatment, whereas surgical treatment yields more favorable early relief of leg pain.

Table 9-2

HELPFUL HINTS

Disk Herniation Classification	MRI Description	Additional Comments
Protrusion	An eccentric bulge within an intact annulus	Often asymptomatic
Extrusion	The diameter of the herniated fragment beyond the base is greater than its diameter at the base of the fragment	These are the most common symptomatic disk herniations
Sequestration	There is no continuity between the herniated fragment and the disc that it originated from	Rare, often referred to as a "free fragment"

REFERENCES

1. Battie MC, Videman T, Parent E. Lumbar disk degeneration: epidemiology and genetic influences. *Spine (Phila Pa 1976)*. 2004;29(23):2679-1690.
2. Videman T, Battie MC, Parent E, et al. Progression and determinants of quantitative magnetic resonance imaging measures of lumbar disk degeneration: a five-year follow-up of adult male monozygotic twins. *Spine (Phila Pa 1976)*. 2008;33(13):1484-1490.
3. Battie MC, Videman T, Gibbons LE, et al. 1995 Volvo Award in clinical sciences: determinants of lumbar disk degeneration: a study relating lifetime exposures and magnetic resonance imaging findings in identical twins. *Spine (Phila Pa 1976)*. 1995;20(24):2601-2612.
4. Sato K, Kikuchi S, Yonezawa T. In vivo intradiskal pressure measurement in healthy individuals and in patients with ongoing back problems. *Spine (Phila Pa 1976)*. 1999;24(23):2468-2474.
5. Fujiwara A, An HS, Lim TH, Haughton VM. Morphologic changes in the lumbar intervertebral foramen due to flexion-extension, lateral bending, and axial rotation: an in vitro anatomic and biomechanical study. *Spine (Phila Pa 1976)*. 2001;26(8)876-882.
6. Rabin A, Gerszten PC, Karausky P, et al. The sensitivity of the seated straight-leg raise test compared with the supine straight-leg raise test in patients presenting with magnetic resonance imaging evidence of lumbar nerve root compression. *Arch Phys Med Rehabil*. 2007;88(7):840-843.

7. Ohshima H, Hirano N, Osada R, et al. Morphologic variation of lumbar posterior longitudinal ligament and the modality of disk herniation. *Spine (Phila Pa 1976)*. 1993;18(16):2408-2411.

8. Forristall RM, Marsh HO, Pay NT. Magnetic resonance imaging and contrast CT of the lumbar spine: comparison of diagnostic methods and correlation with surgical findings. *Spine (Phila Pa 1976)*. 1988;13(9): 1049-1054.

9. Fardon DF, Milette PC; Combined Task Forces of the North American Spine Society, American Society of Spine Radiology, and American Society of Neuroradiology. Nomenclature and classification of lumbar disk pathology: recommendations of the combined Task Forces of the North American Spine Society, American Society of Spine Radiology, and American Society of Neuroradiology. *Spine (Phila Pa 1976)*. 2001;26(5):E93-E113.

10. Schaufele MK, Hatch L, Jones W. Interlaminar versus transforaminal epidural injections for the treatment of symptomatic lumbar intervertebral disk herniations. *Pain Physician*. 2006;9(4):361-366.

11. Vad VB, Bhat AL, Lutz GE, et al. Transforaminal epidural steroid injections in lumbosacral radiculopathy: a prospective randomized study. *Spine (Phila Pa 1976)*. 2002;27(1):11-16.

12. Colonna PC, Fredienburg Z. The disk syndrome. *J Bone Joint Surg Am*. 1949;31A(3):614-618.

13. Bush K, Cowan N, Katz DE, et al. The natural history of sciatica with associated disk pathology: a prospective study with clinical and independent radiologic follow-up. *Spine (Phila Pa 1976)*. 1992;17(10): 1205-1212.

14. Ahn UM, Ahn NU, Buchowski JM, et al. Cauda equina syndrome secondary to lumbar disk herniation: a meta-analysis of surgical outcomes. *Spine (Phila Pa 1976)*. 2000;25(12):1515-1522.

15. Kohles SS, Kohles DA, Karp AP, et al. Time-dependent surgical outcomes following cauda equina syndrome diagnosis: comments on a meta-analysis. *Spine (Phila Pa 1976)*. 2004;29(11):1281-1287.

16. Atlas SJ, Deyo RA, Keller RB, et al. The Maine Lumbar Spine Study, part II: 1-year outcomes of surgical and nonsurgical management of sciatica. *Spine (Phila Pa 1976)*. 1996;21(15):1777-1786.

17. Atlas SJ, Keller RB, Chang Y, et al. Surgical and nonsurgical management of sciatica secondary to a lumbar disk herniation: five-year outcomes from the Maine Lumbar Spine Study. *Spine (Phila Pa 1976)*. 2001;26(10):1179-1187.

18. Atlas SJ, Keller RB, Wu YA, et al. Long-term outcomes of surgical and nonsurgical management of sciatica secondary to a lumbar disk herniation: 10 year results from the Maine Lumbar Spine Study. *Spine (Phila Pa 1976)*. 2005;30(8):927-935.

10

PHYSICAL EXAMINATION FOR LUMBAR SPINAL STENOSIS

Joseph K. Lee, MD and Brian W. Su, MD

INTRODUCTION

Lumbar spinal stenosis is a progressive disorder that is best considered along a spectrum of degenerative changes in the aging spine. Initially described by Portal in 1803, Verbiest popularized the concept of lumbar spinal stenosis in 1949. Although the disorder more commonly causes radicular lower-extremity symptoms, lumbar stenosis can also be a cause of claudicatory low-back pain. The incidence is highest in the fifth decade of life, affecting nearly 1.7% to 8% of the general population.[1] The overall prevalence has increased

Rihn JA, Harris EB. *Musculoskeletal Examination of the Spine: Making the Complex Simple* (pp. 167-182).
© 2011 SLACK Incorporated.

secondary to the increased longevity of the population. There is no gender predominance, and no association with body habitus or occupation.[2]

The natural history of lumbar spinal stenosis is generally benign. Most cases of mild or moderate spinal stenosis do not progress.[3] Johnsson et al[4] observed 32 patients with spinal stenosis treated nonoperatively for 4 years. At follow-up, 70% of patients reported no change in pain level, 15% reported worsening of symptoms, and 15% reported improvement of pain. In the Maine Lumbar Spinal Stenosis study, 67 patients with spinal stenosis underwent follow-up for an average of 4 years. Fifty-two percent of patients reported improvement in symptoms, 18% had worsened leg pain, and 27% had worsened back pain.[5]

HISTORY

Although the constellation of clinical symptoms varies with each case, patients with lumbar spinal stenosis typically present with buttock and leg pain in a distribution of the affected dermatomes.[6] Pathognomonic symptoms include heaviness, cramping, burning, paresthesia, and weakness in the back, buttock, and legs.[7] Patients with foraminal stenosis typically have complaints isolated to a specific nerve root, whereas those with central or lateral recess stenosis may have symptoms that overlap several nerve roots. Prolonged standing, walking, and back extension exacerbate symptoms as those positions close down the neural foramen and central canal. Patients may also indicate activities that promote back extension such as walking down stairs may also worsen symptoms. Sitting or activities such as stationary biking or walking up stairs that promote a leaning forward posture often alleviate symptoms. Although patients may report bowel and bladder dysfunction secondary to pain or narcotic use, true bowel or bladder incontinence is rare with lumbar stenosis.

In a retrospective series of patients with lumbar spinal stenosis, Amundsen et al[8] noted that low-back pain (97%), neurogenic claudication (91%), and leg pain (71%) were the most common reported symptoms. Seventy percent of their patients reported equal severity of symptoms between back and leg complaints, whereas 25% reported radicular symptoms as their primary complaint. The L5 and S1 nerve root were the most commonly affected. Radicular symptoms were unilateral in 58% and bilateral in 42% of cases. Weakness (30%) and voiding disturbances (12%) were relatively uncommon in the population.

EXAMINATION

Most patients with lumbar spinal stenosis may not have overt physical findings. Examination of the lumbar spine begins with inspection of the back and general posture (Table 10-1). Most patients will adopt the "simian posture," where the shoulders are translated anteriorly to the pelvis and the trunk is flexed forward. Patients can compensate for loss of lordosis with hip flexion. Therefore, it is critical to examine patients' posture with the knees locked in extension. Patients may report tenderness to palpation in the lumbosacral region or more often in the sciatic notch where the sciatic nerve exits. Patients also report generalized stiffness associated with decreased range of motion. Symptoms may be exacerbated with forced lumbar hyperextension. Nerve root tension signs such as a straight-leg raise are usually negative unless a concomitant disk herniation is present.

The neurologic examination can be varied in patients with lumbar stenosis. Weakness is often more evident after prolonged walking. All muscle groups from the L2 to S1 distribution should be carefully examined (Figure 10-1). It should be noted that although the tibialis anterior is classically considered as innervated by the L4 nerve root, it is commonly seen to overlap with L5 and may be indicative of L5 nerve compression. True changes in the sensory examination such as vibration, proprioception, and 2-point discrimination are rare. Loss of the patellar (L4) and Achilles reflex (S1) is common, but it should be noted that dampened reflexes in elderly patients is common. Simultaneous pathology in the cervical spine can also present as lower-extremity pathology in patients with lumbar stenosis. It is critical to ask patients if they have problems with balance or fine motor skills and to examine all patients for upper motor neuron compressive signs such as a Hoffmann's sign, Babinski's sign, or clonus (Figure 10-2). If any of these are present with neck pain, cervical magnetic resonance imaging (MRI) is warranted.

Table 10-1

METHODS OF EXAMINING THE BACK FOR LUMBAR SPINAL STENOSIS

EXAMINATION	TECHNIQUE	GRADING	SIGNIFICANCE
Range of motion	Passive and active motion 1. Flexion and Extension 2. Lateral bending 3. Rotation	Degrees	Exacerbation of symptoms with hyperextension, relief with flexion
Palpation	1. Paraspinal muscle 2. Spinous processes 3. Sciatic notch 4. Sacroiliac joint		1. Muscle spasm 2. Evaluate for stepoff in spondylolisthesis 3. Sciatic nerve impingement 4. Sacroilitis
Straight-leg test	Supine on table, lift straight leg upward	Reproduction of radicular symptom	Indicate concomitant disc herniation
Motor exam	1. L2 – Hip flexion via iliopsoas 2. L3 – Knee extension via quadriceps 3. L4 – Ankle dorsiflexion via tibialis anterior 4. L5 – Great toe dorsiflexion via extensor hallucis longus 5. S1 – Ankle plantarflexion via gastrocnemius	Standard motor grading 0-5	Weakness indicates severe or prolonged nerve root compression

(continued)

Table 10-1 (continued)

METHODS OF EXAMINING THE BACK
FOR LUMBAR SPINAL STENOSIS

EXAMINATION	TECHNIQUE	GRADING	SIGNIFICANCE
Pathologic reflexes	1. Babinski's reflex: run sharp instrument across plantar aspect of foot, starting from calcaneus along lateral border of forefoot and ending at the great toe	Positive test- great toe extension with lesser toe flexion and splay	Upper motor neuron injury
	2. Oppenheimer test: run sharp instrument along tibial crest		

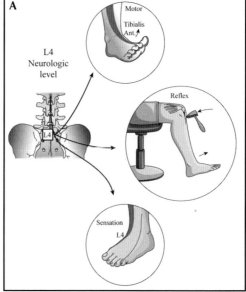

Figure 10-1. The neurologic examinations for assessing (A) L4. *(continued)*

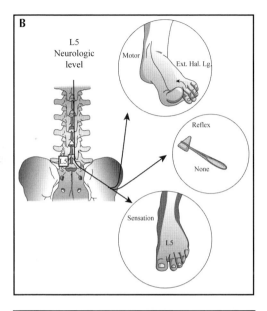

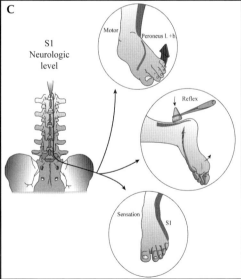

Figure 10-1 (continued). The neurologic examinations for assessing (B) L5 and (C) S1.

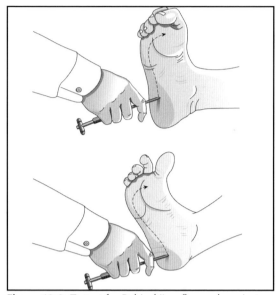

Figure 10-2. To test for Babinski's reflex, a sharp instrument is used to brush across the plantar aspect of the foot, starting from the calcaneus and along the lateral aspect of the foot. A positive reflex occurs with extension of the great toe and splaying of the lesser toes, reflecting an upper motor neuron injury.

Louis and Nazarian[9] reported physical findings from their group of 350 patients and noted that 53% patients had decreased or absent ankle or knee reflex, 37% had objective weakness, and 10% had a positive straight-leg raise. Amundsen et al[8] noted sensory deficits in 51%, reflex abnormalities in 47%, lumbar tenderness in 40%, and weakness in 23%.

PATHOANATOMY

There are several factors that contribute to the development of spinal stenosis. The shape of the spinal canal can predispose an individual to spinal stenosis. The 3 main shapes include round, ovoid, and trefoil (Figure 10-3). The trefoil shape occurs

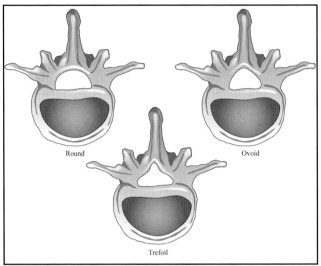

Figure 10-3. Different types of spinal canal shapes include round, ovoid, and trefoil.

in 15% of the population, has the smallest space for the cord, and is present in most cases of symptomatic spinal stenosis.[10] Neural compression from lumbar stenosis typically occurs secondary to arthritic changes at the facet joints, instability, or a congenitally small canal. Stenosis at the lateral recess from spondylotic changes or from a degenerative spondylolisthesis is from the superior articular process of the caudal vertebra compressing the traversing nerve root (ie, stenosis at the L4-L5 level leading to compression of the L5 nerve root by the superior articular process of L5). Foraminal stenosis or an isthmic spondylolisthesis causes compression of the exiting nerve root (ie, stenosis at L4-L5 leading to compression of the L4 nerve root).

The initial insult in the development of spinal stenosis is the degeneration of the intervertebral disk, resulting in the loss of disk height and narrowed vertical height of the neural foramen.[11] Osteophytes and hypertrophic changes of the facet capsule diminish the size of the foramen and lateral recess. Thickened ligamentum flavum can buckle on itself with extension of the spine, further narrowing the size of the canal. Soft-tissue hypertrophy is responsible for 40% of central spinal stenosis.[6] The combination of canal shape and degenerative changes in the bony and soft-tissue elements of the spine contribute to the narrowing of the canal and foraminal space.

IMAGING

Anteroposterior and lateral plain radiographs of the lumbar spine should be obtained in all patients with lumbar spinal stenosis (Figures 10-4A and B). All lateral radiographs should be taken with the knees locked in extension so as not to mask any loss of lumbar lordosis. Flexion-extension lateral radiographs should also be obtained to assess for spondylolisthesis, which may not be evident on a static upright lateral view. It is critical to correlate radiographic pathology with the clinical symptoms, as asymptomatic patients may have significant degenerative changes on their plain radiographs.[12] Loss of disk height, bridging osteophytes between vertebral bodies, and overall alignment of the spine should be noted. Spondylolytic spondylolisthesis, particularly at L5-S1, can be better examined with oblique radiographs of the spine, which better delineates pathology within the pars interarticularis (Figures 10-4C and D).

Advanced imaging such as computed tomography (CT) and MRI are often obtained to provide fine detail of both osseous and soft-tissue structures. MRI will often demonstrate hypertrophy of the facet capsule, thickening and buckling of the ligamentum flavum, and disk herniations (Figure 10-5). It is critical for the surgeon to correlate symptoms with radiographic findings on MRI. Boden et al[13] performed MRI on 67 asymptomatic individuals and noted that in patients >60 years, MRI was positive for herniated disks (57%) and lumbar spinal stenosis (21%).

In patients who are unable to undergo MRI secondary to metal implants, computed tomography myelography (CTM) provides another option. Radiographic dye is injected into the epidural space and a CT scan is performed. Although CTM is an invasive procedure, it provides useful information that can assist with operative planning.

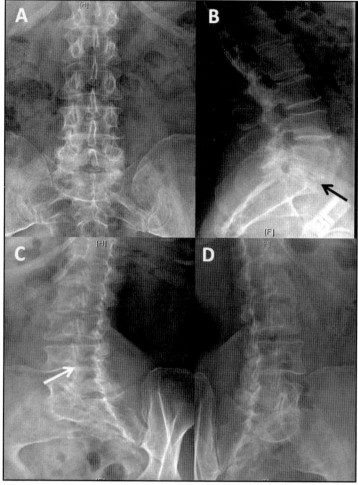

Figure 10-4. Plain radiographs showing lumbar spinal stenosis. (A) AP and (B) lateral radiographs demonstrate loss of disk height at L5-S1 (black arrow), loss of foraminal height, and osteophytes. (C and D) Oblique radiographs are useful for identifying pathology in the pars interarticularis (white arrow).

TREATMENT

Initial treatment should include anti-inflammatory medication, a formal course of physical therapy, and alteration in activities. Strict bed rest is no longer advocated, and patients

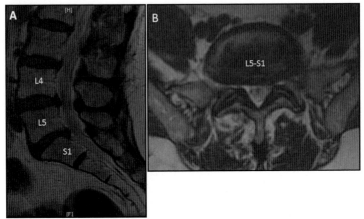

Figure 10-5. (A) Sagittal T2-weighted MRI demonstrates multilevel spinal stenosis at L4-L5 and L5-S1 disk space. (B) Axial T2-weighted MRI demonstrates severe foraminal stenosis secondary to facet hypertrophy, ligamentum flavum hypertrophy, and disk degeneration.

are encouraged to perform activities as soon as they are able to tolerate it. Nonsteroidal anti-inflammatory medications and newer cyclo-oxygenase-2 inhibitors such as Celebrex are useful first-line medications. Oral steroids can be used for short periods to acutely reduce the inflammation of the nerve roots but must be balanced against side effects such as gastritis, hyperglycemia, and potential avascular necrosis of the femoral head. Long-term narcotic use should be avoided, as narcotics do not address the inflammation affecting the nerve roots and are associated with addiction and gastrointestinal problems.

Physical therapy includes flexion exercises, pelvic stabilization, and aerobic exercises. Biomechanical studies of the lumbar spine consistently show that flexion increases the anteroposterior diameter of the canal.[8,14] Cycling is an excellent modality for both weight loss and aerobic conditioning for these patients. The goal of physical therapy is to strengthen the core muscles and allow unloading of the spinal elements. Pelvic traction can be an adjunctive treatment modality. In a study of 51 patients with lumbar spinal stenosis treated with pelvic traction, LaBan and Taylor[15] noted excellent relief in 63%, fair improvement in 27%, and poor response in 10%.

Epidural corticosteroid injections can be considered if prior nonoperative treatment fails to alleviate symptoms. Patients with primary leg complaints may benefit more from epidural injections compared to those with low-back pain as their primary symptom. The efficacy of epidural injections varies. Cuckler et al[16] performed a double-blind, randomized study comparing the efficacy of interlaminar methylprednisolone injection versus a control saline injection. At an average follow-up of 20 months, there was no statistically significant difference between the 2 groups. Rosen et al[17] administered epidural steroid injections in 40 patients with lumbar spinal stenosis. Fifty percent of the patients reported short-term improvement, whereas 25% of the patients reported persistent relief of symptoms. Botwin and Gruber[18] treated 34 patients with spinal stenosis and radicular complaints with an average of 1.9 epidural injections. After 12 months of follow-up, 64% of patients reported better or complete relief of symptoms. Potential complications of epidural injections are rare, with headache being the most common.

Selective nerve root blocks are effective in managing acute radicular leg symptoms in patients with lumbar stenosis. In a prospective, randomized, double-blind study, Riew et al[19] noted that 71% of patients who were scheduled for operative management decided not to have surgery after receiving a selective nerve root block consisting of bupivacaine and betamethasone. Furthermore, a patient's response to a selective nerve root block is a good prognosticator to overall response to surgical decompression. Derby et al[20] noted that 85% of patients with radicular symptoms for >1 year who had relief of symptoms after the selective nerve root block had a good surgical outcome.

Proper surgical indications for lumbar spinal stenosis are important for achieving optimal outcomes. Patients with severe, intractable neurogenic claudication refractory to conservative treatment should be considered for surgery. Isolated back pain without radicular symptoms is not the best indication for surgery, as the results of surgical intervention in this patient group are unpredictable.[21] The goals of operative intervention for spinal stenosis include adequate decompression of the spinal canal and nerve roots while preserving spinal stability. Operative options include laminectomy or selective

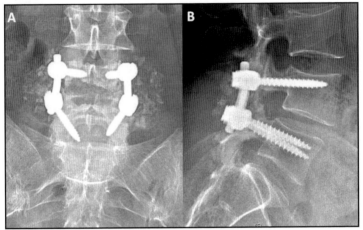

Figure 10-6. Images show laminectomy and posterior segmental instrumentation and fusion for spinal stenosis at L4-L5 level.

laminoforaminotomy. Fusion is performed if there is preexisting instability or if >50% of the facets need to be resected for the decompression (Figure 10-6).

Laminectomy allows full decompression of the central canal and also access to the lateral recesses and foramen. After the appropriate spinal level is determined using intraoperative imaging, the spinous process and lamina are removed with rongeurs and a high-speed burr. The lateral recess and foramen are decompressed using Kerrison rongeurs. It is important to decompress to the medial wall of each pedicle in the mediolateral direction and from the inferior aspect of the superior pedicle to the superior aspect of the inferior pedicle. Care should be taken to avoid injuring the dura or performing an overaggressive decompression that results in iatrogenic instability. At least 50% of the facet complex must remain intact to preserve spinal stability. Laminoforaminotomy focuses on removing the bone and soft tissue that is compressing the neural elements, preserving the posterior spinal elements that are important to stability. This is more commonly performed in unilateral rather than bilateral disease.

Overall results from a wide laminectomy are good, with approximately 60% to 85% of patients reporting good or excellent results from the procedure. In a prospective study of

140 patients with spinal stenosis who underwent laminectomies, Herron et al[22] noted that 80% had significant (>75%) relief of symptoms and an additional 11% had some (25% to 75%) improvement. Complications are relatively low and include nerve root injury, dural tears, iatrogenic instability, and infection.

CONCLUSION

Lumbar spinal stenosis causes neurogenic claudication secondary to narrowing of the spinal canal and neural foramen. When symptoms are severe and refractory to conservative treatment, patients can undergo some type of decompressive procedure and expect good outcomes with relatively low complication rates (Table 10-2).

REFERENCES

1. De Villiers PD, Booysen EL. Fibrous spinal stenosis: a report on 850 myelograms with a water-soluble contrast medium. *Clin OrthopRelat Res.* 1976;(115):140-144.
2. Hilibrand AS, Rand N. Degenerative lumbar stenosis: diagnosis and management. *J Am Acad Orthop Surg.* 1999;7(4):239-249.
3. Postacchini F, Cinotti G, Gumina S, Perugia D. Long-term results of surgery in lumbar stenosis: 8-year review of 64 patients. *Acta Orthop Scand Suppl.* 1993;(251):78-80.
4. Johnsson KE, Rosen I, Uden A. The natural course of lumbar spinal stenosis. *Clin Orthop Relat Res.* 1992;(279):82-86.
5. Atlas SJ, Keller RB, Robson D, Deyo RA, Singer DE. Surgical and nonsurgical management of lumbar spinal stenosis: four-year outcomes from the Maine Lumbar Spine Study. *Spine (Phila Pa 1976).* 2000;25(5): 556-562.
6. Arbit E, Pannullo S. Lumbar stenosis: a clinical review. *Clinical Orthop Relat Res.* 2001;(384):137-143.
7. Garfin SR, Rauschning W. Spinal stenosis. *Instr Course Lect.* 2001;50: 145-152.
8. Amundsen T, Weber H, Lilleas F, Nordal HJ, Abdelnoor M, Magnaes B. Lumbar spinal stenosis: clinical and radiologic features. *Spine (Phila Pa 1976).* 1995;20(10):1178-1186.
9. Louis R, Nazarian S. Lumbar stenosis surgery: the experience of the orthopaedic surgeon. *Chir Organi Mov.* 1992;77(1):23-29.
10. Bolender NF, Schonstrom NS, Spengler DM. Role of computed tomography and myelography in the diagnosis of central spinal stenosis. *J Bone Joint Surg Am.* 1985;67(2):240-246.

Table 10-2

HELPFUL HINTS

Pathophysiology

1. Degenerative disc disease → loss of disc height → loss of height of neural foramen

2. Facet hypertrophy and osteophyte formation → decreased size of foramen/lateral recess

3. Thickened ligamentum flavum → buckles in extension → canal narrowing

IMAGING

Image	Study	Significance
Plain radiographs	1. AP/lateral 2. Oblique 3. Flexion/extension lateral	1. Loss of disc height, bridging osteophytes, overall spinal aligment 2. Lesion in pars interarticularis in setting of sponylolytic spondylolisthesis 3. Presence of spondylolisthesis
CT scan	1. Axial/coronal/sagittal view 2. CT myelogram	1. Evaluate bony detail 2. Use in patients who cannot undergo MRI. Involves injection of dye prior to CT scan. Can evaluate extent of cord/nerve root compression.
MRI	Axial/coronal/sagittal view	1. Evaluate soft tissue detail–hypertrophy of facets, buckling of ligamentum flavum, disc herniation 2. Evaluate severity of cord and nerve root compression

11. Spivak JM. Degenerative lumbar spinal stenosis. *J Bone Joint Surg Am.* 1998;80(7):1053-1066.

12. Frymoyer JW, Newberg A, Pope MH, Wilder DG, Clements J, MacPherson B. Spine radiographs in patients with low-back pain: an epidemiological study in men. *J Bone Joint Surg Am.* 1984;66(7):1048-1055.

13. Boden SD, Davis DO, Dina TS, Patronas NJ, Wiesel SW. Abnormal magnetic-resonance scans of the lumbar spine in asymptomatic subjects: a prospective investigation. *J Bone Joint Surg Am.* 1990;72(3):403-408.

14. Inufusa A, An HS, Lim TH, Hasegawa T, Haughton VM, Nowicki BH. Anatomic changes of the spinal canal and intervertebral foramen associated with flexion-extension movement. *Spine (Phila Pa 1976).* 1996;21(21):2412-2420.

15. LaBan MM, Taylor RS. Manipulation: an objective analysis of the literature. *Orthop Clin North Am.* 1992;23(3):451-459.

16. Cuckler JM, Bernini PA, Wiesel SW, Booth RE Jr, Rothman RH, Pickens GT. The use of epidural steroids in the treatment of lumbar radicular pain: a prospective, randomized, double-blind study. *J Bone Joint Surg Am.* 1985;67(1):63-66.

17. Rosen CD, Kahanovitz N, Bernstein R, Viola K. A retrospective analysis of the efficacy of epidural steroid injections. *Clin Orthop Relat Res.* 1988;(228):270-272.

18. Botwin KP, Gruber RD. Lumbar epidural steroid injections in the patient with lumbar spinal stenosis. *Phys Med Rehabil Clin N Am.* 2003;14(1): 121-141.

19. Riew KD, Yin Y, Gilula L, et al. The effect of nerve-root injections on the need for operative treatment of lumbar radicular pain: a prospective, randomized, controlled, double-blind study. *J Bone Joint Surg Am.* 2000;82(11):1589-1593.

20. Derby R, Kine G, Saal JA, et al. Response to steroid and duration of radicular pain as predictors of surgical outcome. *Spine (Phila Pa 1976).* 1992;17(6 Suppl):S176-S183.

21. Katz JN, Lipson SJ, Brick GW, et al. Clinical correlates of patient satisfaction after laminectomy for degenerative lumbar spinal stenosis. *Spine (Phila Pa 1976).* 1995;20(10):1155-1160.

22. Herron LD, Mangelsdorf C. Lumbar spinal stenosis: results of surgical treatment. *J Spinal Disord.* 1991;4(1):26-33.

11

LUMBAR SPONDYLOLISTHESIS

Ishaq Y. Syed, MD; Justin B. Hohl, MD;
and James D. Kang, MD

INTRODUCTION

The term spondylolisthesis originates from the Greek *spondylos* ("vertebra") and *olisthesis* ("to slip"). It is defined as the anterior translation of one vertebral body relative to an adjacent vertebra. Spondylolysis refers to a dissolution or defect in the pars interarticularis, a narrow area of bone between the superior and inferior articular processes. Wiltse[1] classified spondylolisthesis into 5 types: dysplastic, isthmic, degenerative, traumatic, and pathologic (Table 11-1). The dysplastic and isthmic subtypes are the most common among children and young adults. Dysplastic refers to a congenital abnormality

Rihn JA, Harris EB. *Musculoskeletal Examination of the Spine: Making the Complex Simple* (pp. 183-202).
© 2011 SLACK Incorporated.

Table 11-1

HELPFUL HINTS:
WILTSE SPONDYLOLISTHESIS CLASSIFICATION

Type	Name	Description
I	Dysplastic	Congenital abnormalities of the upper sacrum or posterior arch of L5
II	Isthmic	Lesion of the pars interarticularis; subtypes include A–lytic fatigue fracture of the pars, B–elongated but intact pars, and C–acute fracture
III	Degenerative	Long-standing intersegmental disease
IV	Traumatic	Fracture in regions other than the pars
V	Pathologic	Generalized or local bone disease

of the facets, mostly commonly at L5-S1. The term isthmic, which means "narrow" in Greek, refers to a lesion of the pars interarticularis. There are 3 subclasses of isthmic spondylolisthesis: A–stress fracture of the pars, B–elongation of the pars, and C–acute fracture of the pars. Degenerative spondylolisthesis is a term originally used by Newman and Stone,[2] describing anterior slippage in the presence of an intact neural arch that occurs most often at the L4-L5 level, is most common in the sixth decade of life, and is more common in females than males (6:1).[3] It is thought to be a result of osteoarthritis leading to intersegmental instability of the apophyseal joints and disk degeneration. Traumatic spondylolisthesis is rare and associated with fracture involving the posterior elements. Pathologic spondylolisthesis is associated with generalized bone disease such as neoplasm, osteoporosis, and Paget disease.

The true prevalence of spondylolisthesis is unknown due to a high number of asymptomatic patients.[4] The incidence of spondylolysis in the general population has been reported to be approximately 6% in North America, with twice as many males as females affected, although the risk of progression of slippage appears to be greater in females.[5] The female preponderance is believed to be a result of greater ligamentous laxity and hormonal effects.[6] Certain ethnic groups such as the Eskimo population have a higher incidence of spondylolysis, up to 54% in adults.[7] Race has also been implicated, with a reported incidence of 6.6% in White males versus 2.8% in African American males.[8] Certain athletes including gymnasts, football linemen, divers, weight lifters, and volleyball players have a reported higher incidence as well.[9,10] Spondylolisthesis can be a source of low back pain, radicular symptoms or result in symptoms of neurogenic claudication. A complete history and physical examination are the first steps in diagnosis.

HISTORY

Patients with spondylolisthesis present with a wide range of symptoms, but in the majority of cases, patients remain asymptomatic. The typical clinical presentation of patients with degenerative spondylolisthesis is often the same as spinal stenosis. Patients may present with low-back pain with or without leg pain. Back pain with degenerative spondylolisthesis is typically mechanical, aggravated by extension or rising from a bent posture and relieved by rest. Other sources of back pain must be considered in all patients. Radiation of pain into the posterolateral thighs is common. Leg symptoms can be referred, radicular, or associated with neurogenic claudication. This condition is also termed *pseudoclaudication* and must be distinguished from vascular claudication.

Patients often complain of pain (94%), weakness (43%), numbness (63%), and paresthesias associated with walking or standing.[11] Inquiry about neurologic bowel or bladder incontinence should be part of a complete history, but such symptoms are rare in this disease process. Lower-extremity symptoms consistent with neurogenic claudication typically begin in the buttocks and extend in a proximal-to-distal distribution down the legs. Radicular pain follows a dermatomal distribution, usually L5, and is often unilateral and seen in patients with concomitant lateral recess and foraminal stenosis. Symptoms are often exacerbated in extension, standing, or walking, and resolve with sitting or bending forward. The spinal canal cross-sectional area and foraminal dimension increase in flexion,[12] and therefore, patients may report that they prefer a forward-flexed posture such as leaning on a grocery cart to decrease symptoms. This particular finding has been termed the "shopping cart sign."

Acute spondylolisthesis is usually associated with a significant traumatic event. Patients will complain of pain and disability normally expected from an acute fracture. A child may describe activity-related pain, and only 40% recall a specific traumatic event.[13] The abrupt change in relative stiffness across the lumbosacral junction leaves the pars interarticularis susceptible to fatigue fractures, especially with a repetitive extension moment. Those who do present with symptoms often have back pain and leg pain. The leg pain is predominantly in a dermatomal distribution involving the exiting nerve root caused by lateral recess compression at the level of the pars defect by bony and fibrous tissue elements. As opposed to degenerative spondylolisthesis, patients with the isthmic variety have posterior elements that are dissociated from the vertebral body, and listhesis does not typically result in stenosis of the central canal.

EXAMINATION

Physical examination findings in patients with spondylolisthesis may be nonspecific. The patient's posture may be a fixed forward-flexed position with increased hip and knee flexion termed the Phalen-Dickson sign.[14] A common finding is hamstring tightness with resultant increased popliteal angle on a straight-leg raise. The etiology of hamstring tightness remains unclear, although it is theorized to be secondary to chronic nerve root irritation.[14,15] With increased slippage, the sacrum assumes a more vertical position (flat buttocks appearance), causing the pelvis to flex forward and leading to the clinical appearance of a crouched posture. On examination, the combination of tight hamstrings and a forward-flexed position of the pelvis limits hip extension and causes the knees to flex to maintain sagital balance. The examiner may observe that this causes the patient to ambulate with a pelvic waddle shuffling gait.[16] Hyperlordosis above the level of the slip may also be seen to compensate for the slip. Attempt at forced lumbar hyperextension may reproduce or exacerbate the patient's symptoms.

Table 11-2

METHODS FOR EXAMINATION

Examination	Illustration	Significance
Inspection of overall posture	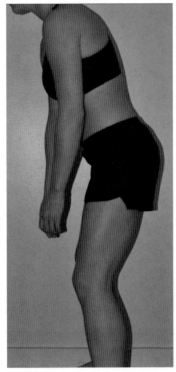	Posture in a fixed, forward-flexed position with increased hip and knee flexion is indicative of a Phalen-Dickson sign; crouched position is more pronounced as the amount of slippage increases

(continued)

Palpation may reveal a step-off of the spinous process above the level of the slip and elicit midline tenderness. Paraspinal muscle spasm may limit flexion in an attempt to prevent shear forces and stabilize the slipped level. Depending on the degree of neural compression, a motor or sensory deficit may be present. Rarely, a patient may present with cauda equina syndrome as the cephalad vertebra, with an intact neural arch, slides forward, effectively narrowing the spinal canal. Table 11-2 presents a detailed review of pertinent examination findings.

Table 11-2 (continued)

METHODS FOR EXAMINATION

Examination	Illustration	Significance
Palpation of spinous processes	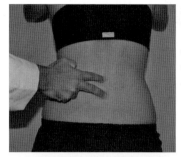	A visible or palpable step-off may indicate forward slippage of one vertebra on another
Palpation of paraspinal muscles	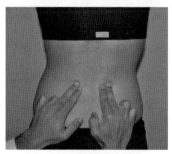	Paraspinal muscles may feel prominent and rigid; unilateral spasm may cause the patient to list to one side and bilateral may obliterate normal lumbar lordosis
Hyperextension of the lumbar spine	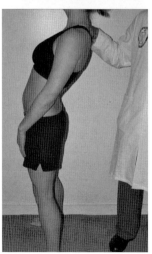	Forced hyperextension elicits pain and reproduce lower-extremity symptoms.

(continued)

Table 11-2 (continued)

METHODS FOR EXAMINATION

Examination	Illustration	Significance
Standing one-legged hyperextension test	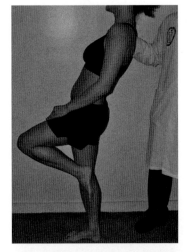	In a single-leg stance on the ipsilateral side in which symptoms are present, the patient leans backward slowly; a positive test reproduces or exacerbates the patient's back pain
Measurement of popliteal angle		With the patient's hip flexed, the pelvis is stabilized, the foot is maintained in dorsiflexion and the knee is extended; the degrees short of full knee extension equals the patients popliteal angle, which is a measure of hamstring tightness

PATHOANATOMY

Degenerative

Development of degenerative spondylolisthesis is dependent on anatomic factors, and progressive degenerative changes of the aging disk can lead to instability.[17] As the disk loses its mechanical integrity, it is unable to resist normal shear forces, allowing the cephalad vertebra to slip anteriorly and subsequently increasing biomechanical forces on the facet joints, leading to arthrosis. More sagittally oriented facet joints as in the lumbar spine (L4-L5)[12,18] or the presence of a pars defect combined with abnormal shear forces allow the vertebral body to translate. The physiologic response to instability is hypertrophy of the facet joints that can help prevent further listhesis but may lead to narrowing of the spinal canal and neuroforamen.

Isthmic

Development of spondylotic defects have an increased incidence with the onset of ambulation and rarely present in nonambulatory patients. This lends support to the role of upright status and repetitive microtrauma in the development of spondylotic defects.[19] Biomechanically, the pars interarticularis has increased shear stresses in extension with persistent lordotic positioning.[20] Biologic attempts at healing the pars defect can result in an increase in cartilage and fibrous tissue, leading to narrowing of the lateral recess and subsequent compression of the exiting nerve root. Certain sporting activities such as gymnastics and football that place the lumbar spine in repetitive extension loading with increased lordosis have been associated with a higher incidence of spondylolysis.[9]

Dysplastic

Genetic predisposition may play a role, especially in the dysplastic type where familial tendency is stronger. Congenital abnormality of the posterior elements including the facet joints are theorized to result in subsequent spondylolisthesis when exposed to stress with increased lumbar lordosis. In the dysplastic type of spondylolisthesis, the pars interarticularis is typically elongated but intact. The association between spina bifida occulta and spondylolisthesis reinforces the theory of congenital basis in some patients (Figure 11-1).[5]

Figure 11-1. Standing lateral radiograph of the lumbar spine showing Meyerding grade IV spondylolisthesis.

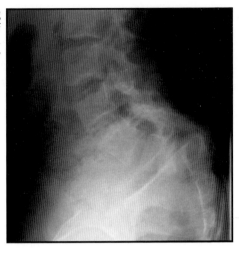

IMAGING

Plain Radiographs

Complete radiographic lumbar series include standing anteroposterior, lateral, bilateral 30 degree oblique, and flexion-extension views. The lateral image is the most valuable in identifying anterolisthesis (Figure 11-2). This view allows grading of the slip (Table 11-3)[21], measurement of the slip angle, and lumbar index. The slip angle is measured by drawing a line perpendicular to the posterior aspect of the sacrum and a line along the inferior end plate of L5. The lateral view may also demonstrate degenerative changes including loss of disk height, osteophyte formation, and facet arthrosis. Oblique radiographic views may allow more clear visualization of pars interarticularis defects. This area is classically referred to as the neck of the "Scotty dog" (Figure 11-3). The AP radiographs may demonstrate an associated scoliosis or spina bifida occulta. Flexion-extension lateral radiographs are valuable in visualizing dynamic instability of a listhetic segment or identifying subtle slips. Dynamic change >4 mm is considered abnormal.[22] In patients who have the appropriate clinical presentation but in whom plain radiographs are inconclusive, more advanced imaging may be appropriate.

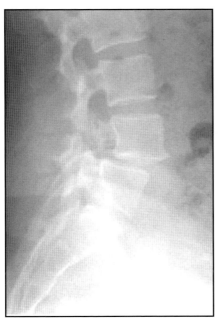

Figure 11-2. Standing lateral radiograph of the lumbar spine demonstrating isthmic spondylolisthesis of L4 on L5.

Table 11-3

MEYERDING CLASSIFICATION

Grade	% Slip
I	0 to 25
II	26 to 50
III	51 to 75
IV	76 to 100
V (Spondyloptosis)	>100

Radionuclide

Single-photon emission computed tomography (SPECT) has a higher sensitivity and provides more detail than a standard bone scan for suspected acute pars lesions.[23] An increased

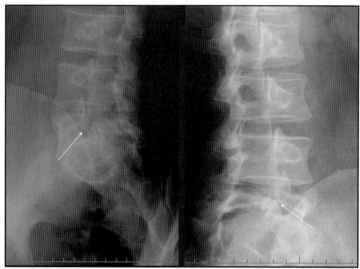

Figure 11-3. Arrows on the oblique radiographs point to bilateral lysis of the pars interarticularis, commonly referred to as a break in the neck of the "Scotty dog" (eye = pedicle, nose = transverse process, ear = superior articular process, foreleg = inferior articular process, neck = pars interarticularis, and body = lamina).

uptake in the suspected area indicates increased metabolic activity in the region of an acute lesion that reverts to normal as the lysis becomes chronic with or without healing (Figure 11-4). SPECT scan intensity varies with time and stability in spondylolysis.[24] Increased uptake may indicate healing potential and may help guide formulation of an appropriate treatment plan.

Computed Tomography

Computed tomography (CT) is the study of choice in demonstrating detailed bony architecture. It can be used to further delineate the presence of a pars defect with associated sclerosis, demonstrate degenerative changes in the apophyseal joints, and assess stenosis[25] (Figure 11-5). CT has been shown to be more sensitive than magnetic resonance imaging (MRI) and standard bone scans and can be used to monitor healing of spondylolytic defects.[26] The downside is significant radiation exposure, which may be of concern especially in the pediatric population. Concomitant myelogram may

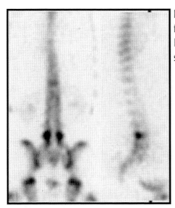

Figure 11-4. Bone scan shows intense focal activity over the pedicles of the L4 vertebra consistent with bilateral L4 spondylolysis.

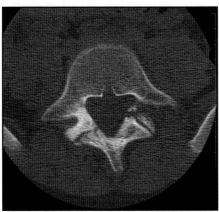

Figure 11-5. Axial CT showing the presence of bilateral fibrocartilaginous pars defect and associated sclerosis leading to lateral recess stenosis.

play an important role in evaluating patients for compression of neural elements, especially if MRI is contraindicated.

Magnetic Resonance Imaging

MRI has become the gold standard in evaluating soft-tissue abnormalities associated with disks, ligaments, and neural structures. It provides excellent anatomic detail in assessing the degree of impingement of neural elements both in the central canal and foramina (Figures 11-6, 11-7, and 11-8). In addition, MRI allows evaluation of associated degenerative changes in the disks and facet joints.[27] It may also demonstrate a "wide canal sign" suggestive of bilateral pars defect. Findings must be carefully correlated to the patient's signs and symptoms. The clinical relevance of disk abnormalities visualized

Figure 11-6. Midsagittal T2-weighted MRI demonstrating isthmic spondylolisthesis of L4 on L5. Note the relative disk degeneration and focal canal stenosis at L4-L5.

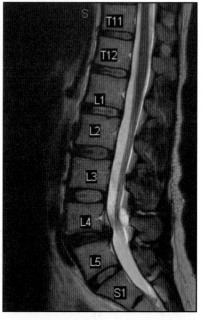

Figure 11-7. Sagittal T1-weighted MRI showing foraminal narrowing as a result of listhesis of the L4-L5 level causing impingement of the L4 exiting nerve root.

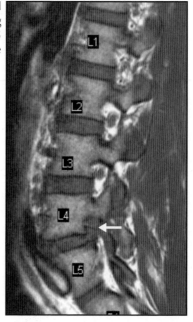

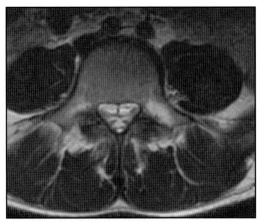

Figure 11-8. Axial T2-weighted MRI demonstrating bilateral fibrocartilaginous pars defect. Note lytic lesions are present on the same cut as the pedicles, a hallmark finding in correctly identifying pars defect.

on MRI associated with spondylolisthesis is unclear. An MRI study performed in the supine position may fail to identify a listhesis preoperatively; therefore, standing lateral lumbar radiographs should be obtained in addition to MRI. Due to its increasing availability, superior anatomic detail, and lack of ionizing radiation, MRI has largely replaced CT myleogram in evaluating the neural elements.

TREATMENT

Conservative

Nonoperative treatment in adults can include short-term relative rest (1 to 2 weeks), judicious use of nonsteroidal anti-inflammatory drugs, and a short course of oral steroids that is best reserved for acute exacerbations of leg pain. Narcotic analgesics should be avoided but may be used sparingly for short durations in cases of acute exacerbation of pain. Physiotherapy programs that focus on abdominal core strengthening and hip flexor/hamstring stretching are also widely utilized. Flexion-based physical therapy is generally considered superior to extension-based programs in providing symptomatic relief.[28]

There is no clear evidence to support the use of physical therapy modalities such as heat, ultrasound, and massage, although these modalities may provide temporary symptomatic relief for some patients.

Patients should be encouraged to incorporate a daily exercise routine for maintaining overall aerobic conditioning, muscular strength, and flexibility. The next line of treatment may be referral to pain management for epidural steroid injections. In the short term this may provide relief of radicular symptoms but is unlikely to improve back pain. Conservative nonsurgical treatment in neurologically normal patients has been demonstrated to result in satisfactory clinical outcome at an average of 10-year follow-up in the majority of patients.[3]

In skeletally immature patients, the general treatment is nonoperative in patients presenting with <50% slip. Most commonly, a teenage athlete with associated repetitive trauma in a lordotic position may present with symptoms. Activity modification and avoidance of sports is the initial treatment of choice.[29] Children treated nonoperatively have resolution of symptoms >80% of the time, with healing of the acute lesions estimated to be between 75% and 100%.[30]

Bracing may have a role in patients who continue to be symptomatic despite activity modification or who have progression of slippage. Low-profile underarm plastic braces have shown good to excellent results in up to 80% of patients with grade 1 and 2 spondylolisthesis.[31] In acute injury, spondylolysis appears to respond favorably to rapid initiation of bracing, which leads to more rapid return to activities and fewer patients with persistent discomfort.[23] Typically, patients are allowed to resume full activity on resolution of symptoms and of no evidence of progression during brace treatment.[31] Those with higher-grade deformity are less likely to respond to nonoperative management.[32]

Surgical

No specific guidelines have been well established for surgical intervention in adults with spondylolisthesis. Patients who fail to respond to conservative measures lasting 6 months or longer, have progressive neurologic deficit, persistent or recurrent severe leg pain, or neurogenic claudication causing significant disability may be candidates for surgery. An absolute indication for surgery is the development of cauda equina syndrome; though this is rarely associated with spondylolisthesis other than in high-energy trauma cases. The goal of surgical treatment is stabilization of the affected levels and decompression of the neural elements. Stabilization by fusion can result in improvement and resolution of neural deficit by preventing continued motion and irritation of neural tissues.[33]

The recent Spine Patient Outcomes Research Trial reported that surgical treatment in patients with degenerative lumbar spondylolisthesis with associated spinal stenosis yielded greater pain relief and improved function at 4-year follow-up compared to nonoperative treatment.[34] Surgical treatment options include decompression without fusion, decompression with noninstrumented posterolateral fusion, decompression with instrumented posterolateral fusion, and concomitant posterior fusion with anterior column support. The optimal surgical treatment remains controversial. It is generally accepted that patients with degenerative spondylolisthesis have a better result with fusion compared to decompression alone,[35] instrumentation appears to increase the rate of fusion,[36] and patients with a solid fusion have improved long-term clinical results.[37]

Loose posterior elements are removed along with any hypertrophic fibrous tissue at the pars defect in adults with isthmic spondylolisthesis. Decompression alone has been associated with poor outcomes, accelerated disk degeneration, and up to 27% slip progression rate at long term follow-up.[38] An overall clinical success rate of 75% has been reported in patients undergoing posterior fusion for isthmic spondylolisthesis.[39] Improved fusion rate (90% versus 77%) and clinical success (85% versus 64%) has been demonstrated with the addition of instrumentation.[39] Although controversial, some advocate the addition of anterior column support to provide indirect decompression, increase surface area for fusion, provide correction of sagittal alignment, and maintain reduction.[40] Various techniques for interbody fusion exist including anterior lumbar interbody fusion, posterior lumbar interbody fusion, transforaminal lumbar interbody fusion, and, more recently, transposes direct lateral interbody fusion. Statistically significant improvement in clinical scores in circumferential fusion has been demonstrated over posterolateral fusion alone, with nonsignificant trends toward improvement in fusion rates[41] (Figure 11-9).

Indications for surgical stabilization in skeletally immature patients include recalcitrant nonoperative treatment, grade III

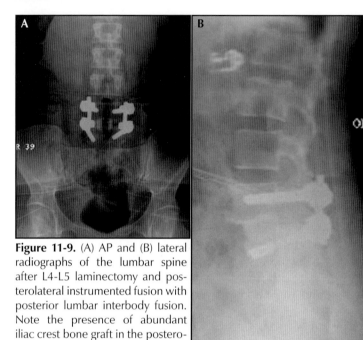

Figure 11-9. (A) AP and (B) lateral radiographs of the lumbar spine after L4-L5 laminectomy and posterolateral instrumented fusion with posterior lumbar interbody fusion. Note the presence of abundant iliac crest bone graft in the posterolateral gutters on the AP view and restoration of the foraminal height on the lateral view.

or greater slip, and progressive deformity.[42] The gold standard for surgical management includes posterior spinal fusion from L5 to S1 without reduction and extending the fusion proximally to L4 if there is 50% or greater anterior translation.[43] The addition of anterior column support may provide indirect decompression of neural elements, an improved rate of fusion, and better restoration of sagittal alignment.[44] Direct repair of the pars interarticularis defect with bone grafting may be a valid option in patients with minimal degrees of slippage in whom symptoms persists despite nonoperative management. Using a tension-band construct has been reported to result in 80% good to excellent results and 90% radiographic healing,[45] although various types of fixation have been described in the literature. In long-term analysis, no statistically significant difference in outcome has been shown in patients treated with direct pars repair compared to posterolateral fusion with similar rates of adjacent-segment degeneration in the long term.[46]

CONCLUSION

Lumbar spondylolisthesis is a common condition, and careful assessment with complete history, physical examination, and imaging studies is critical to establish an accurate diagnosis. The majority of patients can successfully be treated with nonsurgical management and have symptomatic relief. In appropriately selected symptomatic patients who remain recalcitrant to nonsurgical treatment, surgical management may be considered. The specific technique is controversial and should be individualized for each patient based on clinical and imaging findings to maximize patient outcome. Further prospective randomized studies are needed to determine the optimal technique given the specific disease pattern.

REFERENCES

1. Wiltse LL, Newman PH, MacNab I. Classification of spondylolisis and spondylolisthesis. *Clin Orthop Relat Res.* 1976;(117):23-29.
2. Newman P, Stone KH. The etiology of spondylolisthesis. *J Bone Joint Surg Br.* 1963;45(1):39-59.
3. Matsunaga S, Sakou T, Morizono Y, Masuda A, Demirtas AM. Natural history of degenerative spondylolisthesis: pathogenesis and natural course of the slippage. *Spine (Phila Pa 1976).* 1990;15(11):1204-1210.
4. Vaccaro AR, Martyak GG, Madigan L. Adult isthmic spondylolisthesis. *Orthopedics.* 2001;24(12):1172-1177.
5. Fredrickson BE, Baker D, McHolick WJ, Yuan HA, Lubicky JP. The natural history of spondylolysis and spondylolisthesis. *J Bone Joint Surg Am.* 1984;66(5):699-707.
6. Bird HA, Eastmond CJ, Hudson A, et al. Is generalized joint laxity a factor in spondylolisthesis? *Scand J Rheumatol.* 1980;9(4):203-205.
7. Simper LB. Spondylolysis in Eskimo skeletons. *Acta Orthop Scand.* 1986;57(1):78-80.
8. Rowe GG, Roche MB. The etiology of separate neural arch. *J Bone Joint Surg Am.* 1953;35(1):102-110.
9. Semon RL, Spengler D. Significance of lumbar spondylolysis in college football players. *Spine (Phila Pa 1976).* 1981;6(2):172-174.
10. Jackson DW, Wiltse LL, Cirincoine RJ. Spondylolysis in the female gymnast. *Clin Orthop Relat Res.* 1976;(117):68-73.
11. Katz JN, Dalgas M, Stucki G, et al. Degenerative lumbar spinal stenosis: diagnostic value of the history and physical examination. *Arthritis Rheum.* 1995;38(9):1236-1241.
12. Inufusa A, An HS, Lim TH, Hasegawa T, Haughton VM, Nowicki BH. Anatomic changes of the spinal canal and intervertebral foramen associated with flexion-extension movement. *Spine (Phila Pa 1976).* 1996;21(21):2412-2420.
13. El Rassi G, Takemitsu M, Woratanarat P, Shah SA. Lumbar spondylolysis in pediatric and adolescent soccer players. *Am J Sports Med.* 2005;33(11):1688-1693.

14. Phalen GS, Dickson JA. Spondylolisthesis and tight hamstrings. *J Bone Joint Surg Am.* 1961;43(4):505-512.
15. Deyerle WM. Lumbar-nerve-root irritation in children. *Clin Orthop.* 1961;(21):125-136.
16. Meyers LL, Dobson SR, Wiegand D, Webb JD, Mencio GA. Mechanical instability as a cause of gait disturbance in high-grade spondylolisthesis: a pre- and postoperative three-dimensional gait analysis. *J Pediatr Orthop.* S 1999;19(5):672-676.
17. Rosenberg NJ. Degenerative spondylolisthesis: surgical treatment. *Clin Orthop Relat Res.* 1976;(117):112-120.
18. Grobler LJ, Robertson PA, Novotny JE, Pope MH. Etiology of spondylolisthesis: assessment of the role played by lumbar facet joint morphology. *Spine (Phila Pa 1976).* 1993;18(1):80-91.
19. Rosenberg NJ, Bargar WL, Friedman B. The incidence of spondylolysis and spondylolisthesis in nonambulatory patients. *Spine (Phila Pa 1976).* 1981;6(1):35-38.
20. Letts M, Smallman T, Afanasiev R, Gouw G. Fracture of the pars interarticularis in adolescent athletes: a clinical-biomechanical analysis. *J Pediatr Orthop.* 1986;6(1):40-46.
21. H M. Spondylolisthesis. *Surg Gynecol Obstret.* 1932;(54):371-377.
22. Boden SD, Wiesel SW. Lumbosacral segmental motion in normal individuals: have we been measuring instability properly? *Spine (Phila Pa 1976).* 1990;15(6):571-576.
23. Anderson K, Sarwark JF, Conway JJ, Logue ES, Schafer MF. Quantitative assessment with SPECT imaging of stress injuries of the pars interarticularis and response to bracing. *J Pediatr Orthop.* 2000;20(1):28-33.
24. Lusins JO, Elting JJ, Cicoria AD, Goldsmith SJ. SPECT evaluation of lumbar spondylolysis and spondylolisthesis. *Spine (Phila Pa 1976).* 1994;19(5):608-612.
25. Teplick JG, Laffey PA, Berman A, Haskin ME. Diagnosis and evaluation of spondylolisthesis and/or spondylolysis on axial CT. *AJNR Am J Neuroradiol.* 1986;7(3):479-491.
26. Saifuddin A, Burnett SJ. The value of lumbar spine MRI in the assessment of the pars interarticularis. *Clin Radiol.* 1997;52(9):666-671.
27. Birch JG, Herring JA, Maravilla KR. Splitting of the intervertebral disc in spondylolisthesis: a magnetic resonance imaging finding in two cases. *J Pediatr Orthop.* 1986;6(5):609-611.
28. O'Sullivan PB, Phyty GD, Twomey LT, Allison GT. Evaluation of specific stabilizing exercise in the treatment of chronic low back pain with radiologic diagnosis of spondylolysis or spondylolisthesis. *Spine (Phila Pa 1976).* 1997;22(24):2959-2967.
29. Seitsalo S, Osterman K, Hyvarinen H, Tallroth K, Schlenzka D, Poussa M. Progression of spondylolisthesis in children and adolescents: a long-term follow-up of 272 patients. *Spine (Phila Pa 1976).* 1991;16(4):417-421.
30. Sys J, Michielsen J, Bracke P, Martens M, Verstreken J. Nonoperative treatment of active spondylolysis in elite athletes with normal X-ray findings: literature review and results of conservative treatment. *Eur Spine J.* 2001;10(6):498-504.
31. Bell DF, Ehrlich MG, Zaleske DJ. Brace treatment for symptomatic spondylolisthesis. *Clin Orthop Relat Res.* 1988;(236):192-198.

32. Frennered AK, Danielson BI, Nachemson AL, Nordwall AB. Midterm follow-up of young patients fused in situ for spondylolisthesis. *Spine (Phila Pa 1976)*. 1991;16(4):409-416.

33. LL W. Spondylolisthesis and its treatment. In: *Low Back Pain*. 2nd ed. Philadelphia, PA: JB Lippincott; 1980:451-493.

34. Weinstein JN, Lurie JD, Tosteson TD, et al. Surgical compared with non-operative treatment for lumbar degenerative spondylolisthesis: four-year results in the Spine Patient Outcomes Research Trial (SPORT) randomized and observational cohorts. *J Bone Joint Surg Am*. 2009;91(6):1295-1304.

35. Herkowitz HN, Kurz LT. Degenerative lumbar spondylolisthesis with spinal stenosis: a prospective study comparing decompression with decompression and intertransverse process arthrodesis. *J Bone Joint Surg Am*. 1991;73(6):802-808.

36. Fischgrund JS, Mackay M, Herkowitz HN, Brower R, Montgomery DM, Kurz LT. 1997 Volvo Award winner in clinical studies: degenerative lumbar spondylolisthesis with spinal stenosis: a prospective, randomized study comparing decompressive laminectomy and arthrodesis with and without spinal instrumentation. *Spine (Phila Pa 1976)*. 1997;22(24):2807-2812.

37. Kornblum MB, Fischgrund JS, Herkowitz HN, Abraham DA, Berkower DL, Ditkoff JS. Degenerative lumbar spondylolisthesis with spinal stenosis: a prospective long-term study comparing fusion and pseudarthrosis. *Spine (Phila Pa 1976)*. 2004;29(7):726-733.

38. Amuso SJ, Neef RS, Coulson DB, Laing PG. The surgical treatment of spondylolisthesis by posterior element resection: a long-term follow-up study. *J Bone Joint Surg Am*. 1970;52(3):529-536.

39. Kown BK, Hilibrand AS, Malloy K, et al. A critical analysis of the literature regarding surgical approach and outcome for adult low-grade isthmic spondylolisthesis. *J Spinal Disord Tech*. 2005;(18)(suppl):S30-S40.

40. Ekman P, Moller H, Tullberg T, Neumann P, Hedlund R. Posterior lumbar interbody fusion versus posterolateral fusion in adult isthmic spondylolisthesis. *Spine (Phila Pa 1976)*. 2007;32(20):2178-2183.

41. Swan J, Hurwitx E, Malek F, et al. Surgical treatment for unstable low-grade isthmic spondylolisthesis in adults: a prospective controlled study of posterior instrumented fusion compared with combined anterior-posterior fusion. *Spine J*. 2006;6(6):606-614.

42. Boxall D, Bradford DS, Winter RB, Moe JH. Management of severe spondylolisthesis in children and adolescents. *J Bone Joint Surg Am*. 1979;61(4):479-495.

43. Poussa M, Schlenzka D, Seitsalo S, Ylikoski M, Hurri H, Osterman K. Surgical treatment of severe isthmic spondylolisthesis in adolescents: reduction or fusion in situ. *Spine (Phila Pa 1976)*. 1993;18(7):894-901.

44. Kawakami M, Tamaki T, Ando M, Yamada H, Hashizume H, Yoshida M. Lumbar sagittal balance influences the clinical outcome after decompression and posterolateral spinal fusion for degenerative lumbar spondylolisthesis. *Spine (Phila Pa 1976)*. 2002;27(1):59-64.

45. Bradford DS, Iza J. Repair of the defect in spondylolysis or minimal degrees of spondylolisthesis by segmental wire fixation and bone grafting. *Spine (Phila Pa 1976)*. 1985;10(7):673-679.

46. Schlenzka D, Seitsalo S, Poussa M, et al. Operative treatment of symptomatic lumbar spondylolysis and mild spondylolisthesis in young patients: direct repair of the defect or segmental spinal fusion. *Eur Spine J*. 1993;2(2):104-112.

12

LUMBAR
DEGENERATIVE
DISK DISEASE

Vidyadhar V. Upasani, MD; Steven R. Garfin, MD;
and R. Todd Allen, MD, PhD

INTRODUCTION

Axial lumbar pain (ie, low back pain) is one of the most common musculoskeletal complaints in the general population. It has been estimated that nearly 80% of individuals are affected by this symptom at some point in their lives.[1] Although low-back pain is often attributed to disk degeneration, the actual degenerative process is a normal part of aging and does not always cause pain. The natural history of lumbar degenerative disk disease is one of recurrent episodes of pain followed by periods of significant or complete relief.[2] The exact etiology of this pain, however, is often difficult to accurately diagnose. Although the presence of radiating pain into one or

Rihn JA, Harris EB. *Musculoskeletal Examination of the Spine: Making the Complex Simple* (pp. 203-222).
© 2011 SLACK Incorporated.

Table 12-1

HELPFUL HINTS FOR HISTORY AND IMAGING OF LUMBAR DISK DEGENERATION

History

- Lumbar disk degeneration is an inevitable consequence of aging and is frequently asymptomatic.

- Symptoms need to be correlated with the physical examination and imaging findings for correct diagnosis.

- Low-back pain may be exacerbated by repetitive twisting activities, prolonged sitting, or exposure to constant vibration.

- Watch for red flags that may indicate an infection, a neoplasm, or a pathologic fracture.

- Loss of perineal sensation, severe pain, leg weakness, and loss of voluntary bowel or bladder control may indicate cauda equina syndrome.

Imaging

- Radiographs (AP, lateral, oblique, and flexion-extension views) can evaluate for spinal alignment, instability, disk space narrowing, vertebral endplate sclerosis, osteophytes, and facet arthropathy.

- MRI is the imaging study of choice to detect intraspinal pathology but has high sensitivity and may detect asymptomatic pathology.

- CTM is useful to detect stenosis and herniated disks if MRI is contraindicated.

more extremities, especially in a dermatomal distribution, can greatly aid in localizing the pathology, most patients present with radiographic evidence of multilevel degenerative changes. Regardless, a thorough history and physical examination, supported by confirmatory imaging studies, should be performed in each individual to determine the most appropriate surgical or nonsurgical treatment (Table 12-1).

The primary focus of the initial medical evaluation for axial lumbar pain should be to rule out a variety of etiologies that would require more urgent diagnostic evaluation and management. For example, traumatic sources of back pain could include compression or burst fractures, soft-tissue sprains or strains, or traction neuritis. Neoplasms including metastatic disease (most commonly from a primary site of the breast, lung, or prostate) or primary spinal cord or bone tumors should be considered. Infectious etiologies could include diskitis, osteomyelitis, soft-tissue or epidural abscess, or tuberculosis. Inflammatory conditions such as seronegative spondyloarthropathies (eg, ankylosing spondylitis and Reiter syndrome) or rheumatoid arthritis, and neurologic conditions such as demyelinating disease, anterior horn cell disease, and intraspinal cysts could cause low-back pain. Finally, rarer causes to consider include visceral conditions (ie, abdominal and renal problems, inflammatory bowel disease, or pelvis disease), cardiac and pulmonary illnesses, herpes zoster, polymyalgia rheumatica, myofascial syndromes, diffuse idiopathic skeletal hyperostosis, cauda equina syndrome, Paget, and psychogenic etiologies.

HISTORY

Several risk factors have been associated with symptomatic lumbar degenerative disk disease including obesity, cigarette smoking, sedentary lifestyle, and male gender.[3] Certain occupational activities may also predispose these patients to axial low back pain. These include repetitive lifting or prolonged exposure to vibrations such as with driving long distances. These patients most commonly present with lumbar pain of a mechanical nature that is often exacerbated by heavy exertion, repetitive bending, or twisting. Furthermore, diskogenic low-back pain is typically worse with sitting and improved with standing and walking. In advanced degenerative states with multilevel stenosis, patients may present with extremity pain and fatigue that limits ambulation. Patients often report that these symptoms improve with rest or forward flexion (classic findings of neurogenic claudication).

Although some patients may identify a specific traumatic incident related to the onset of their symptoms, most report intermittent episodes of back pain that have been present for many months or even years. Patients usually describe their pain as starting in the lower back or through the sacroiliac region. Some may describe the pain as band-like or may describe radiating symptoms into the buttocks or posterior thighs. This should be distinguished from radicular-type pain (sciatica), which usually extends below the knees and typically follows a dermatomal distribution of the affected nerve root(s).

A detailed history should be used to rule out other diagnoses in the differential diagnoses. For example, pain at rest or pain that wakes the patient from sleep should be evaluated for an infection, neoplasm or pathologic fracture. Prolonged stiffness that improves with activity may indicate an inflammatory condition. A nonanatomic distribution of symptoms, generalized gait difficulties, or strength deficits should raise concern for intracranial or peripheral neurologic disorders, metastatic conditions, or cervical or thoracic spine pathology, which can coexist in a substantial percentage of patients. Most importantly, the history should also rule out symptoms of cauda equina syndrome including loss of bowel or bladder control and perianal or genital dysesthesia.

EXAMINATION

Physical examination findings (Table 12-2) in patients with degenerative disk disease may vary based on the waxing and waning time course of the condition.[4] The use of a pain drawing or pain and function tests (eg, Oswestry disability index or SF 12/36) can provide helpful objective information in the evaluation and treatment of patients with chronic low back pain.

When patients first present, they should be observed for their general behavior and postural changes that may be attributable to axial back or leg pain. An important starting point is to evaluate spinal alignment in both the coronal and sagittal planes. Patients in acute pain may develop paraspinal muscle spasms resulting in a scoliotic deformity (coronal) or a loss of normal lumbar lordosis, or even kyphosis (sagittal). If they present with leg pain, muscle atrophy in the affected limb may be appreciated and can be documented by

Table 12-2

METHODS FOR EXAMINING THE LUMBAR SPINE

Observation

- Fluidity of movement and gait
- Skin markings, masses, or hairy patch
- Posture and spinal alignment
- Muscle atrophy

Palpation

- Paraspinal muscle spasm or tenderness
- Sciatic notch tenderness in radiculopathy
- Step-off in spinous processes indicative for spondylolisthesis

Range of motion

- Usually decreased secondary to pain
- Pain with extension indicative of posterior element pathology (facet arthritis)
- Pain with flexion indicative of diskogenic pain or radiculopathy

Neurologic examination

- T12, L1, L2, L3
  - Motor to iliopsoas, quadriceps, and adductors
  - Sensation to anterior and mid-thigh
- L4
  - Motor to tibialis anterior
  - Sensation to medial side of leg
  - Patellar reflex
- L5
  - Motor to extensor hallucis longus
  - Sensation to dorsum of foot

(continued)

Table 12-2 (continued)

METHODS FOR EXAMINING THE LUMBAR SPINE

- S1

  - Motor to peroneus longus and brevis, and gastrocnemius-soleus complex

  - Sensation to lateral side, plantar foot, and small toe

  - Achilles reflex

Pathologic reflexes

- Babinski's test: Stimulation to plantar and lateral border of foot results in extension of great toe in a positive test

- Oppenheim test: Stimulation along crest of tibia results in extension of great toe in a positive test

Provocative tests

- Straight-leg raise test: Differentiates between stretch of sciatic nerve versus hamstring tightness by dorsiflexion of foot at near extreme and symptoms in foot

- Femoral nerve stretch test: Provokes radicular pain (L2 to L4 distribution) by extending hip with knee in flexion and patient prone

measuring the circumference of the thigh or calf (muscular atrophy in a myotomal distribution may be an indication of neural compression). Also, observation of gait is important to document any loss of coordination or the presence of a slow, unsteady gait, which can be seen with myelopathy, Parkinson disease, or intracranial pathologies.

Paraspinal muscle spasms may be palpable, and although they may be intermittent, they can be intensely painful and tender to the touch. Occasionally palpation of the vertebral bodies may elicit severe pain, inconsistent with the magnitude of force used. A step-off between the spinous processes may indicate the presence of a spondylolisthesis. Sciatic notch tenderness to palpation may also be appreciated, most notably along the course of the sciatic nerve in patients with radiculopathy.

Pain related to percussion or palpation over the sacroiliac joint may be useful as an etiology of the symptoms. Various physical examination maneuvers have been described to evaluate the sacroiliac joint. The best known is the Patrick test (or fabere sign) that involves flexion, abduction, external rotation, and extension of the involved extremity to elicit pain. Another commonly used maneuver is the Gaenslen test whereby the involved hip is hyperextended and patients describe pain adjacent to the posterosuperior iliac spine. Finally, a diagnostic injection in the sacroiliac joint guided by fluoroscopy or computed tomography (CT) can be helpful to rule out sacroiliac pathology.

Lumbar spine range of motion (flexion, extension, lateral bending, and rotation) may be variable but is often decreased secondary to pain. Painful and dysrhythmic range of motion can be indicative of mechanical spinal instability but is not reliable. Pain at only the extremes of motion may also be a helpful finding. The exacerbation of low back pain in extension may be suggestive of posterior element pathology (consistent with facet arthropathy and degenerative spinal stenosis), whereas worsening symptoms with flexion (which loads the intervertebral disks) may be indicative of diskogenic pain. The onset of leg pain with lumbar flexion is usually the result of increased nerve tension and suggests radiculopathy due to a lumbar disk herniation. With a posterolateral lumbar disk herniation associated with degenerative disk disease, lateral bending may cause pain in the ipsilateral limb as the nerve root is stretched over the herniation. On the other hand, in patients with axillary herniated disks, lateral bending away from the symptomatic side will stretch the nerve root, resulting in worsening symptoms.

Patients with isolated axial back pain usually have normal neurologic examinations; however, in patients with radiculopathy, a focused motor, sensory, and reflex neurologic examination should be performed to localize the level of root involvement (Figure 12-1). In general, nerve root dysfunction results in muscle weakness, hyporeflexia, and occasionally muscle atrophy. For example, L4 nerve root compression can result in pain and numbness localized around the anteromedial aspect of the knee and leg, motor weakness in the tibialis anterior muscle (+/- quadriceps) demonstrated on strength

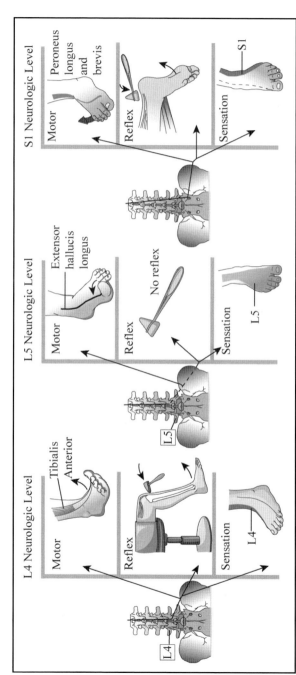

Figure 12-1. Neurologic evaluation of the lower extremity, L4 to S1 levels.

testing or heel-walk, and a diminished or absent patellar tendon reflex. L5 radiculopathy can cause pain on the antero-lateral aspect of the upper leg crossing over to the dorsum of the foot and great toe, with motor weakness in the extensor hallucis longus muscle. S1 nerve root findings can cause pain on the lateral and plantar aspect of the foot and small toe with weakness in the gastrocnemius-soleus complex and a reduced or absent ankle jerk reflex. Symptoms of myelopathy including hyperreflexia, muscle spasticity (positive clonus) and muscle weakness may also be present, often suggesting spinal cord compression in the cervical or thoracic spine.

Certain provocative tests can be used to elicit symptoms of radiculopathy in patients with lumbar disk herniation. For example, the Lasegue sign or straight-leg raise test in either the sitting or supine position should reproduce the patient's radicular pain in a specific dermatomal pattern (sciatic nerve). Isolated back pain with straight-leg raise is not a positive finding; however, it may be indicative of a symptomatic annular tear. Radicular symptoms in the contralateral leg with straight-leg raise may suggest the presence of a sequestered or large extruded disk fragment within the spinal canal. The femoral nerve stretch test is performed with the patient prone. Extension of the leg with the knee flexed may irritate the femoral nerve, resulting in radicular symptoms in the L2 to L4 dermatomal distribution. Finally, Waddell nonorganic tests are important to identify patients who may be malingering. These include nonanatomic superficial tenderness, stimulation tests (axial loading and rotation), flip test (straight-leg raise positive while supine but negative while sitting), nonanatomic weakness and sensory findings, and overreaction.[5]

PATHOANATOMY

Spinal column degeneration can be thought of as a normal aging process that occurs in 3 separate stages.[6] During the first stage, circumferential and radial tears in the intervertebral disk annulus and localized synovitis of the facet joints occur as a result of mechanical overloading of the spine, as well as a genetic predisposition to degeneration. In the next stage, progression to internal disruption of the disk and facet joint

erosion and subluxation may result in disk bulging, herniation, and possibly spinal instability. In the final stage, progressive arthrosis with development of so-called stabilizing osteophytes and enlargement of facet articular processes results in
multilevel spondylosis with central canal, lateral recess, and
foraminal stenosis, as well as a relative increase in segmental
spinal stiffness. Throughout this degenerative process, the
pain generators that result in symptomatic dysfunction and
disability have been studied; however, they need to be more
clearly identified. It is still unknown why certain patients
report significant pain with relatively benign imaging studies,
whereas other patients can have evidence of multilevel degeneration and be asymptomatic. Nevertheless, current treatment
modalities attempt to provide symptomatic relief while not
actually reversing or curing the degenerative process.

The 2 primary pathologic sites that are most commonly
addressed with modern treatment strategies are the intervertebral disks and the facet joints (Figure 12-2). Intervertebral
disks consist of an outer annulus fibrosus which is anchored
to the vertebral body, and an inner nucleus pulposus. The
nucleus pulposus consists of negatively charged proteoglycans
that attract water, giving the disk turgor and height, and
the ability to resist compressive loads. Motion segment
degeneration is thought to begin with a loss of water and
proteoglycan content in the nucleus pulposus as well as a
disruption of collagen crosslinks in the annulus fibrosus.
As the disk continues to degenerate, there is progressive
replacement of Type II collagen with Type I collagen primarily in the nucleus pulposus, and tears or fissures develop
in the annulus fibrosus, resulting in progressive ingrowth
of vascular tissue and nociceptive fibers. Concurrently,
the vertebral body end plates calcify, which impairs nutri-
ent diffusion to the center of the disk, further accelerating
degeneration. Eventually, the nucleus pulposus may extrude
from the center of the disk, resulting in nerve root irritation
or compression.

As the degenerative process progresses in the anterior
spine, facet joints in the posterior spine can also be affected.
Once again, it is unclear if the disk and facet joints degenerate
concurrently or if there is a direct cause and effect relationship
between these 2 processes. The facet capsules initially develop
synovitis, which is often followed by increasing capsular laxity

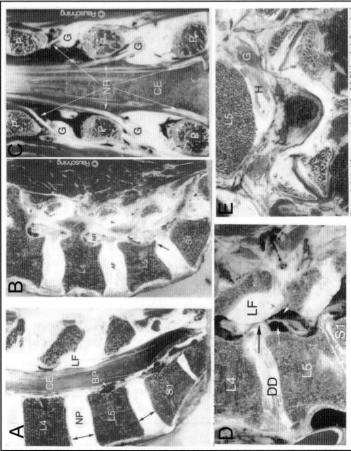

Figure 12-2. Normal and degenerative lumbar spine anatomy. (A) Midsagittal, (B) parasagittal, and (C) coronal views show a normal lumbar spine. Note the maintained disk heights and nerve roots exiting below the pedicles and out the neural foramen. The coronal section shows the cauda equina and the location of nerve root ganglia in relation to the pedicle. (D) Sagittal view shows a degenerative lumbar segment. Notice the narrowed degenerative disk, posterior disk bulge, ligamentum flavum hypertrophy, and narrowed intervertebral foramen for exiting the nerve root. (E) Axial view shows a herniated disk. (Abbreviations: AF, annulus fibrosus; BP, below pedicle; F, facet; CE, cauda equina; DD, degenerative disk; G, nerve root ganglia; H, herniated disk; LF, ligamentum flavum; NP, nucleus pulposus; NR, nerve roots; P, pedicle; reprinted with permission of Wolfgang Rauschning, MD, PhD.)

and hypermobility. These degenerative changes may predispose to various patterns of segmental instability and manifest as spondylolisthesis or degenerative scoliosis. Eventually, the development of osteoarthrosis in these joints results in joint space narrowing, subchondral sclerosis, subchondral cysts, and osteophyte formation. Coupled with thickening of the ligamentum flavum, the degenerative process can lead to narrowing of the spinal canal and neural compression (central, lateral recess, or foraminal).

IMAGING

The diagnosis of lumbar degenerative disk disease should be based primarily on the history and physical examination as >80% of asymptomatic individuals >55 years have evidence of degeneration on imaging studies.[7] In the setting of acute low back pain, imaging studies are typically not performed unless the symptoms persist longer than 4 to 6 weeks or unless certain red flags are present (eg, night pain, constitutional symptoms, or progressive neurologic deficit). Most episodes of acute low back pain will resolve within 4 to 6 weeks with conservative treatment.

When evaluating patients with low back pain, the initial imaging studies should always be anteroposterior and lateral plain radiographs to evaluate the coronal and sagittal spinal alignment, and specifically assess for disk space narrowing, vertebral end-plate sclerosis, facet arthropathy, and osteophyte formation. The radiographs can also be useful to rule out spondylolysis, spondylolisthesis, and diffuse idiopathic skeletal hyperostosis, as well as the presence of fractures, tumors, or inflammatory arthropathies. Lateral flexion-extension radiographs can be used to assess for instability, defined as >4.5 mm of translation or >15 to 20 degrees of angulation between adjacent vertebrae.

CT, computed tomography myelography (CTM), and magnetic resonance imaging (MRI) are commonly used for diagnosis and preoperative planning. Bell et al found myelography to be more accurate than CT for identifying herniated nucleus pulposus and spinal stenosis,[8] whereas Szypryt et al found

MRI to be more accurate than myelography for detecting spinal abnormalities.[9] CT is most useful for evaluating ossified posterior longitudinal ligaments, foraminal stenosis, and bone changes. Although CTM is invasive and increases the risk of arachnoiditis, it can be necessary when MRI is unavailable or contraindicated due to the presence of a cardiac pacemaker or metallic devices.

MRI is the imaging test of choice to evaluate for neural compression and assess disk appearance; however, studies have not been able to show a strong correlation between the severity of findings on MRI and clinical symptoms.[7] The nucleus pulposus, in its healthy state, consists mostly of water which appears as high signal intensity on the T2-weighted MRI sequence. Degenerative disk disease is seen on the T2-weighted MRI sequence as a "blackening" of the disk. The degenerative disk appears black on this MRI sequence because the nucleus pulposus loses its water content as it degenerates. Additional evidence of degenerative disk disease, including annular tearing as well as vertebral body end-plate changes and vertebral body edema can be seen on MRI. Gadolinium contrast can be useful with MRI to better differentiate complex lesions such as tumors, infections, and recurrent herniated disks (versus scarring, which is typically more vascular).

Other diagnostic studies can be useful to rule out diseases other than primary disk herniation, spinal stenosis, and spinal arthritis. Electromyography and nerve conduction studies, for example, can identify peripheral neuropathy and diffuse neurologic involvement indicating higher or lower neural lesions. Nuclear bone scan can confirm neoplastic, traumatic, or arthritic problems of the spine.

Diskography can play an important role in the diagnosis of degenerative disk disease. The indications for this procedure include surgical planning, identifying a symptomatic disk among multiple degenerative disks, and evaluating the structural integrity of disks adjacent to a spondylolisthesis or fusion. When diskography elicits concordant pain in the same anatomic distribution as the patient's symptoms, the study provides information regarding the clinical significance of the disk abnormality and may be useful in identifying the pain generator.

TREATMENT

Nonsurgical Treatment

Various nonoperative treatment modalities exist for patients with lumbar degenerative disk disease. These modalities should be the primary options in patients with axial back pain without neurologic dysfunction. The simplest treatment for acute back pain is short periods of rest[10] (<2 days) with ice or heat therapy, physical therapy, and nonsteroidal anti-inflammatory medications. Short-term oral steroids, opiate analgesics, or muscle relaxants can sometimes be helpful; however, it must be reinforced that chronic use of these medications can result in drug habituation and depression. As pain control improves, patients should be encouraged to begin isometric exercises and low-impact cardiovascular activities.[11] Physical therapy can also play an important role, not only to teach core strengthening exercises but also for education in proper posture and body mechanics.[12]

Alternative methods are available, such as ultrasound, herbal remedies, traction, and electrical nerve stimulation. Although low–level, nonrandomized studies or anecdotal reports suggest that these modalities can provide limited benefit, there are no high-level or longer-term studies supporting their use at this time. Finally, some high-level evidence does exist suggesting that acupuncture may be beneficial in this population, but long-term data are lacking, and the control group in these studies may not reflect patients undergoing other nonoperative treatment modalities for lumbar disk disease.

Spinal injections may provide an additional nonoperative modality to treat leg pain.[13] A recent systematic review demonstrated that transforaminal epidural injections reduced leg pain by 64% to 81%, disability by 60%, and depression by 56%.[14] However, these injections were not found to prevent subsequent surgery, and their benefits were limited to 6 weeks to 3 months. The underlying mechanism of action of these injections is thought to be an alteration or interruption of nociceptive input, the reflex mechanism of afferent fibers, and the pattern of central neuronal activation achieved by the neural blockade. The addition of corticosteroids has been shown to reduce inflammation; however, long-lasting benefits in terms of pain control have been demonstrated with local anesthetics alone in animal models.[15] In the posterior spinal elements, a recent review demonstrated moderate evidence for short-term back pain relief with intra-articular facet joint injections, medial branch blocks, and neurotomy[16] (radiofrequency ablation or cryoneurolysis); however, the long-term success rate was low.

Surgical Treatment

Patients requiring surgical treatment for lumbar degenerative disk disease can be divided into 2 groups: emergent and elective. In the surgical treatment of lumbar degenerative disk disease, one must discriminate the stenosis that occurs as a result of a long-term degenerative process (Figure 12-3) from the stenosis that occurs with more acutely compressive disk pathologies, such as large lumbar disk herniations (Figure 12-4), which can result in severe neural compression and cause cauda equina dysfunction or progressive motor and sensory deficits. Indications for emergent surgery include profound or progressive neurologic deficit or cauda equina syndrome. For all other patients (ie, elective), a comprehensive nonoperative program using the treatments described above should be considered and tried for at least 6 to 12 weeks. If the patient experiences minimal benefit or develops worsening pain and disability, surgical interventions to decompress and stabilize the spine may be necessary.

Several strategies exist to decompress the neural elements using a posterior approach, including foraminotomy, laminotomy, or complete laminectomy. These procedures are intended to treat radicular leg symptoms or neurogenic claudication caused by lumbar disk herniation or lumbar stenosis but are not intended to treat diskogenic low back pain. In terms of treating axial back pain, only 4 Level I randomized controlled studies have compared lumbar fusion surgery to nonoperative treatment for lumbar degenerative disk disease and axial low-back pain. Although some studies show significant benefits with fusion, results should be interpreted with caution due to the variability of surgical interventions used, study heterogeneity, nonstandardized control populations, and variable inclusion and exclusion criteria. Alternatively, lumbar artificial disk replacements have been used to maintain spinal motion after a decompression procedure. Although short-term data demonstrate similar results compared to fusion,[17,18] nonindustry-sponsored trials are needed to adequately assess the

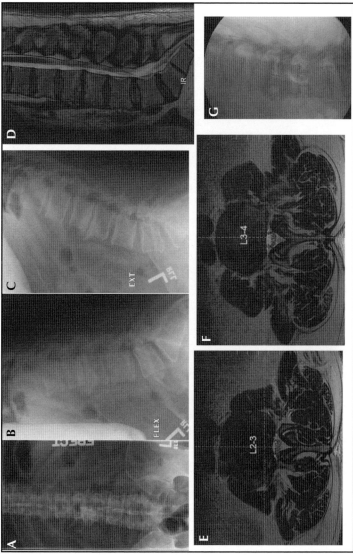

Figure 12-3. (A) Anteroposterior, (B) flexion lateral, and (C) extension lateral radiographs. (D) T2-weighted sagittal, (E) axial of L2-3, and (F) axial of L3-4 MRIs of a 61-year-old man with a 30+ year history of low back pain and progressively worsening intermittent bilateral lower extremity pain. He was diagnosed with degenerative disc disease of L2-3 and L3-4 and underwent an anterior and posterior decompression and fusion procedure. (G) Intra-operative lateral radiograph.

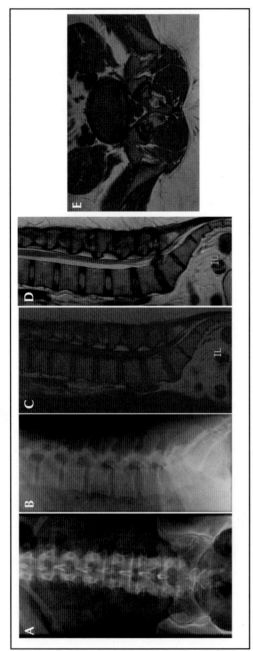

Figure 12-4. (A) Anteroposterior and (B) lateral lumbar spine radiographs. (C) Sagittal T2-weighted, (D) sagittal T1-weighted, and (E) axial MRIs of a 37-year-old woman who developed acute left lower-extremity pain, numbness, and weakness after a 6-hour airplane trip. She was diagnosed with a central L5-S1 disk herniation and underwent a left sided L5-S1 microdiskectomy after failing an 8-week course of nonsurgical management with physical therapy and transforaminal injections.

long-term risks and benefits, as several complications relating to device failure and need for revision surgery have been reported.

There is no clear evidence regarding the most effective technique of decompression for spinal stenosis or the benefit provided by an adjunct fusion (with or without instrumentation). Especially when treating chronic nonradicular low-back pain with surgical decompression and fusion, inconsistent results have been reported, with only 16% of patients experiencing an "excellent" outcome with no pain, restriction of function, or need for analgesics.[19] When radicular symptoms are present, standard open or microscopic diskectomies have been shown to be more effective than nonoperative modalities. Although some data suggested these benefits may decrease over time,[20,21] data (Level I + II data) from the Spine Patient Outcomes Research Trial have shown that surgical treatment of lumbar disk herniation is significantly better than nonsurgical treatment in all measures of leg and back pain 4 years postoperatively.[22] As these procedures are electively performed in a majority of patients with lumbar disk disease, shared decision making should be used to carefully review the potential benefits, harms, costs, and burdens of the surgical procedure for each individual patient.[23]

CONCLUSION

Lumbar disk degeneration is a normal part of the aging process; however, it can result in debilitating low-back pain in some patients. The natural history of this condition is usually benign, with a majority of patients experiencing significant improvement in pain and function after a period of 4 to 6 weeks. A thorough history and physical examination is important to identify red flags that can be indicative of a more aggressive pathologic process. Imaging studies should be used to confirm the diagnosis and for preoperative planning. Numerous nonoperative modalities may provide benefit; however, in patients with progressive symptoms or neurologic deficits, surgical decompression and fusion may be indicated. The benefit of lumbar fusion or total disk arthroplasty over nonoperative treatment for diskogenic low-back pain is not yet proven.

REFERENCES

1. Hult L. The Munkfors investigation: a study of the frequency and causes of the stiff neck-brachialgia and lumbago-sciatica syndromes, as well as observations on certain signs and symptoms from the dorsal spine and the joints of the extremities in industrial and forest workers. *Acta Orthop Scand Suppl.* 1954;16:1-76.

2. Grauer JN, Beiner JM, Albert TJ. Lumbar disc disease. In: Vacarro AR, ed. *Orthopaedic Knowledge Update 8: Home Study Syllabus.* Rosemont, IL: American Academy of Orthopaedic Surgeons; 2005:539-552.

3. Andersson GB. Epidemiological features of chronic low-back pain. *Lancet.* 1999;354(9178):581-585.

4. Haak M. History and physical examination. In: Spivak J, Connolly P, eds. *Orthopaedic Knowledge Update Spine 3.* Rosemont, IL: American Academy of Orthopaedic Surgeons; 2006:43-56.

5. Waddell G, McCulloch JA, Kummel E, Venner RM. Nonorganic physical signs in low-back pain. *Spine (Phila Pa 1976).* 1980;5(2):117-125.

6. Rao R, Bagaria V. Pathophysiology of degenerative disk disease and related symptoms. In: Spivak J, Connolly P, eds. *Orthopaedic Knowledge Update Spine 3.* Rosemont, IL: American Academy of Orthopaedic Surgeons; 2006:35-42.

7. Videman T, Battie MC, Gibbons LE, Maravilla K, Manninen H, Kaprio J. Associations between back pain history and lumbar MRI findings. *Spine (Phila Pa 1976).* 2003;8(6):582-588.

8. Bell GR, Rothman RH, Booth RE, et. A study of computer-assisted tomography, II: comparison of metrizamide myelography and computed tomography in the diagnosis of herniated lumbar disc and spinal stenosis. *Spine (Phila Pa 1976).* 1984;9(6):552-556.

9. Szypryt EP, Twinning P, Wilde GP, Mulholland RC, Worthington BS. Diagnosis of lumbar disc protrusion: a comparison between magnetic resonance imaging and radiculopathy. *J Bone Joint Surg Br.* 1988;70(5):717-722.

10. Deyo RA, Diehl AK, Rosenthal M. How many days of bed rest for acute low back pain: a randomized clinical trial. *N Engl J Med.* 1989;315(17):1064-1070.

11. Malmivaara A, Hakkinen U, Aro T, et al. The treatment of acute low back pain—bed rest, exercise or ordinary activity? *N Engl J Med.* 1995;332(6):351-355.

12. Bergquist-Ullman M, Larsson U. Acute low back pain in industry: a controlled prospective study with special reference to therapy and confounding factors. *Acta Orthop Scand.* 1977;(170):1-117.

13. Buttermann GR. Treatment of lumbar disc herniation: epidural steroid injection compared with discectomy: a prospective, randomized controlled study. *J Bone Joint Surg Am.* 2004;86(4):670-679.

14. Buenaventura RM, Datta S, Abdi S, Smith HS. Systematic review of therapeutic lumbar transforaminal epidural steroid injections. *Pain Physician.* 2009;12(1):233-251.

15. Tachihara H, Sekiguchi M, Kikuchi S, Konno S. Do corticosteroids produce additional benefit in nerve root infiltration for lumbar disc herniation? *Spine (Phila Pa 1976).* 2008;33(7):743-747.

16. Boswell MV, Colson JD, Sehgal N, Dunbar EE, Epter R. A systematic review of therapeutic facet joint interventions in chronic spinal pain. *Pain Physician.* 2007;10(1):229-253.

17. Blumenthal S, McAfee PC, Guyer RD, et al. A prospective, randomized, multicenter Food and Drug Administration investigational device exemptions study of lumbar total disc replacement with the CHARITE artificial disc versus lumbar fusion, Part I: evaluation of clinical outcomes. *Spine (Phila Pa 1976).* 2005;30(14):1565-1575.

18. Zigler J, Delamarter R, Spivak JM, et al. Results of the prospective, randomized multicenter Food and Drug Administration investigational device exemption study of the ProDisc-L total disc replacement versus circumferential fusion for the treatment of 1-level degenerative disc disease. *Spine (Phila Pa 1976).* 2007;32(11):1155-1162.

19. Fritzell P, Hagg O, Wessberg P, Nordwall A. Volvo award winner in clinical studies: lumbar fusion versus Nonsurgical treatment for chronic low back pain: A multicentre randomized controlled trial from the Swedish Lumbar Study Group. *Spine (Phila Pa 1976).* 2001;26(23): 2521-2534.

20. Peul WC, van Houwelingen HC, van den Hout WB, et al. Surgery versus prolonged conservative treatment for sciatica. *N Engl J Med.* 2007;356(22):2245-2256.

21. Weinstein JN, Tosteson TD, Lurie JD, et al. Surgical vs nonoperative treatment for lumbar disk herniation: the Spine Patient Outcomes Research Trial (SPORT): a randomized trial. *JAMA.* 2006;296(20): 2441-2450.

22. Weinstein JN, Lurie JD, Toseteson TD, et al. Surgical compared with nonoperative treatment for lumbar degenerative spondylolisthesis; four-year results in the Spine Patient Outcomes Research Trial (SPORT) randomized and observational cohorts. *J Bone Joint Surg Am.* 2009;91(6): 1295-1304.

23. Chou R, Baisden J, Carragee EJ, Resnick DK, Shaffer WO, Loeser JD. Surgery for low back pain; a review of the evidence for an American Pain Society Clinical Practice guideline. *Spine (Phila Pa 1976).* 2009;34(10):1094-1109.

13

THORACOLUMBAR
TRAUMA

*Jeffrey A. Rihn, MD; Harvey E. Smith, MD;
and Alexander R. Vaccaro, MD, PhD*

INTRODUCTION

Trauma to the thoracolumbar spine is usually the result of a high-energy injury, such as a motor vehicle accident or a fall from height. Concomitant multisystem injuries, including injury to the gastrointestinal tract, are relatively common in thoracolumbar injury patients. The presence of a thoracolumbar injury should prompt a thorough evaluation in an emergency care setting. This includes a careful primary and secondary survey in the emergency room. Great care should be taken to maintain spinal precautions during the evaluation and resuscitation. The secondary survey should include a thorough neurologic examination. Imaging studies should include plain

Rihn JA, Harris EB. *Musculoskeletal Examination of
the Spine: Making the Complex Simple* (pp. 223-241).
© 2011 SLACK Incorporated.

radiographs, computed tomography (CT), and, in many cases, magnetic resonance imaging (MRI). Radiographs and CT will show the bony injuries, whereas MRI is particularly helpful in demonstrating soft-tissue injury, including injury to the intervertebral disk, posterior spinal ligaments, and spinal cord.

The classification and management of thoracolumbar injuries are topics of debate. Numerous classification systems have been described in the past. Most recently, the comprehensive, reproducible classification system of the Spine Trauma Study Group has prognostic significance and can guide treatment decisions. The Thoracolumbar Injury Classification and Severity Score (TLICS) classifies thoracolumbar injures based on 3 pivotal characteristics: the morphology of the injury, the integrity of the posterior ligamentous complex, and the patient's neurologic status.[1] A total severity score is used in conjunction with the classification system to determine the treatment. The majority of injuries can be managed nonoperatively with a brace; however, some injuries require surgical management due to the presence of instability and/or neurologic compromise.

HISTORY

Taking a thorough history of a thoracolumbar spine trauma patient is often challenging. Nonetheless, an effort should be made to gain as much information as possible regarding the mechanism of injury, the demographics and medical status of the patient, and any information pertaining to the neurologic status of the patient at the scene of the trauma. This is particularly important if the patient requires intubation in the "field" prior to arrival in the emergency room. In this situation, the history must be obtained from observers and/or members of the on-site medical emergency team (eg, emergency medical technicians or paramedics).

Understanding the mechanism of injury is important, as this information can give insight into the pattern of thoracolumbar injury. Information regarding the type of trauma (eg, motor vehicle accident or fall from a height) also is important. In the case of a motor vehicle accident, it is important to know where the person was sitting, whether the person was wearing a seatbelt, and the type of accident that occurred (ie, head-on, rear-end, or broad-side collision). As an example, a common

Table 13-1

METHODS FOR EXAMINATION

Primary trauma survey	Airway
	Breathing
	Circulation
Secondary trauma survey	Spine
	Chest
	Abdomen
	Pelvis
	Extremities
Full neurological examination	Motor (upper and lower extremities)
	Sensory (upper and lower extremities)
	Rectal exam
	Reflexes:
	○ Bulbocavernosus
	○ Deep tendon

pattern of thoracolumbar fracture is that of a flexion-distraction injury, which commonly occurs in seatbelted patients who are involved in a high-energy, head-on collision. This injury has a high incidence of concurrent injury to the gastrointestinal tract. Knowledge of this mechanism will facilitate appropriate evaluation and management.

PHYSICAL EXAMINATION

The physical examination of the thoracolumbar injury patient should begin with the primary survey, which includes assessment of the airway, breathing, and circulation (ABCs; Table 13-1). A patent airway should be confirmed or established, and it should be confirmed that the patient is breathing. Patients who are not breathing sufficiently should be immediately intubated and ventilated. Hemodynamic instability should be addressed through fluid resuscitation and by treating any contributing factors, such as tension pneumothorax. The level of consciousness of the patient should be assessed using the Glasgow Coma Scale, which measures eye-opening, verbal and motor-behavioral responses.

After the primary survey is completed and the patient is stabilized, the secondary survey should be performed. The secondary survey represents an evaluation for evidence of injury to the spine, head, chest, abdomen, pelvis, and extremities. During the secondary survey, the cervical spine should be stabilized and the patient should be turned in a controlled fashion to allow examination of the spinal column. The skin overlying the posterior thoracic and lumbar spine should be examined for evidence of laceration, abrasion, swelling, or ecchymosis. The entire posterior spine should then be palpated and examined for point-tenderness, step-off, and spaying of the spinous processes.

A complete neurologic examination should be performed in thoracolumbar trauma patients as part of the secondary survey. The American Spinal Injury Association (ASIA) Impairment Scale (AIS) is used to assess the level of neurologic disability of the patient (Table 13-2).[2,3] The AIS provides information that is helpful in determining the prognosis of the patient. The neurologic examination should include a complete motor and sensory (light touch and pinprick) examination of the upper and lower extremities, evaluation of the upper- and lower-extremity deep tendon reflexes, and in the setting of spinal cord injury, an attempt to localize the level and extent of injury.

In a spinal cord injury patient, the motor level of neurologic injury is defined as the most caudal motor level (eg, C7 motor level determined by testing the triceps muscle) that tests a 3 of 5 on manual motor testing, as long as the next most cranial level tests a full 5 of 5 on manual motor testing. The sensory level of neurologic injury is defined as the most caudal sensory level (ie, according to the sensory dermatomes) at which the patient has intact sensation to light touch and pinprick. A rectal examination is performed as part of the neurologic evaluation, checking for rectal tone, perianal sensation, volition, and the bulbocavernosus reflex. If a patient is unresponsive, the neurologic evaluation should test for the presence of spontaneous movement or movement in response to painful stimuli, as well as the presence or absence of deep tendon reflexes, the bulbocavernosus reflex, and rectal tone.[4]

Table 13-2

AMERICAN SPINAL INJURY ASSOCIATION IMPAIRMENT SCALE

AIS Grade	Neurologic Injury	Description of Neurologic Status
A	Complete	No motor or sensory function below the level of neurologic injury; no motor or sensory function preserved in sacral segments S4-S5
B	Incomplete	Sensory but not motor function preserved below the level of neurologic injury and includes the sacral segments S4-S5
C	Incomplete	Motor function preserved below the level of neurologic injury; more than half of the key muscles below the level of neurologic injury have a muscle grade <3 of 5
D	Incomplete	Motor function preserved below the level of neurologic injury; more than half of the key muscles below the level of neurologic injury have a muscle grade ≥3 of 5
E	Normal	Motor and sensory function are normal

Adapted from Marino RJ, Barros T, Biering-Sorenson F, et al. International standards for neurological classification of spinal cord injury. *J Spinal Cord Med.* 2003;26 (suppl 1):S50-S56.

When evaluating a thoracolumbar injury patient, it is important to understand the concepts of neurogenic and spinal shock. Neurogenic shock is manifested by hypotension in the setting of bradycardia and normal urine output. This is in contrast to hemorrhagic shock, which results from the loss of intravascular volume (blood loss) and is manifested by hypotension, tachycardia, and decreased urine output. Neurogenic shock is typically secondary to spinal cord injury in the cervical or upper thoracic spine rather than the thoracolumbar spine.[5] Spinal shock is different from neurogenic shock and refers to the acute depression of spinal reflexes (eg, bulbocavernosus reflex) distal to the level of injury that results from a physiologic disruption within the acutely injured spinal cord. Spinal shock typically lasts 24 to 48 hours after the spinal cord injury. Return of the bulbocavernosus reflex is indicative of resolution of spinal shock.[6]. The presence of a complete spinal cord injury cannot be confirmed until spinal shock resolves.

IMAGING STUDIES

In the setting of acute thoracolumbar trauma, patients with back pain, thoracolumbar tenderness or step-off on examination, altered mental status, a distracting injury, and a high-energy mechanism of injury should undergo plain radiographs and CT. An MRI should also be obtained if there is any evidence of neurologic injury or suspicion of injury to the soft-tissue elements of the thoracolumbar spine.[7-9] Imaging is only obtained after the patient has been stabilized.

Anteroposterior and lateral thoracolumbar radiographs are typically obtained first, although many trauma centers are moving toward obtaining CT as part of the initial work-up. Although significant fractures and dislocations are usually visible on the plain radiographs, CT provides much more detail of the bony elements of the spine and can more accurately diagnose thoracolumbar spine fractures, subluxation, and dislocation. Axial CT images combined with sagittal and coronal reconstructions can identify most bony injuries. The use of CT in the setting of acute spinal trauma has become ubiquitous in trauma centers and in many trauma centers has replaced plain radiography as the first imaging study for thoracolumbar injury patients.

MRI should be obtained in the setting of acute thoracolumbar trauma in patients who have any neurologic symptoms or signs. It is also obtained to assess the intervertebral disk and posterior ligamentous complex of the thoracolumbar spine for injury. This is particularly helpful in patients with ankylosing spondylitis (AS) or diffuse idiopathic skeletal hyperostosis (DISH). In patients with these disorders, fractures are common in the setting of even low-energy trauma, such as a ground level fall. Even in the setting of normal plain radiographs and CT, patients with AS or DISH with any degree of trauma who complain of back pain should undergo MRI to rule out an occult fracture.

CLASSIFICATION OF
THORACOLUMBAR INJURY

Injury classification systems serve to facilitate accurate communication between clinicians of fracture characteristics, accurate fracture classification for research purposes, and fracture classification to guide treatment decisions. Consequently, a classification system should have good inter- and intra-observer reproducibility and should convey the salient clinical characteristics of the injury. Advances in diagnostic imaging have led to greater understanding of fracture characteristics and mechanisms of injury, and the incorporation of assessment of the associated soft-tissue structures in fracture classification. Attempts to classify spine trauma injury have been complicated, difficult to recall, do not take into account the neurologic status of the patient, and have a great deal of variety among clinicians. Thoracolumbar injuries have traditionally been classified by the 2-column (Holdsworth),[10], 3-column (Denis),[11] Ferguson/Allen,[12] or AO classification systems.[13]

2-Column Classification System

The 2-column system is composed of the anterior column and the posterior column. The anterior column consists of the anterior longitudinal ligament (ALL), vertebral body, and the posterior longitudinal ligament (PLL). The posterior column consists of the vertebral arch, facet joints, and the spinous processes. In the 2-column system, the posterior column is considered most critical to spinal stability.[10]

3-Column Classification System

Based on a review of 412 cases, Denis[11] proposed the 3-column system. The anterior column consists of the ALL and the anterior two-thirds of the vertebral body. The middle column consists of the posterior one-third of the vertebral body and the PLL. The posterior column consists of all structures posterior to the PLL (Figure 13-1). Within this 3-column system, the middle column is felt to be paramount to spinal stability. Injury of 2 or more columns is necessary to create a situation that is unstable enough to require surgical treatment. Both the 2- and 3-column classification systems are based on anatomy rather than injury mechanism or the neurologic status of the patient.

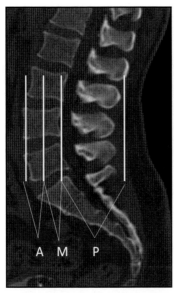

Figure 13-1. Sagittal CT of the thoracolumbar spine demonstrating the 3 columns of the spine according to Denis' 3-column classification system. The anterior column consists of the anterior longitudinal ligament and anterior two-thirds of the vertebral body. The middle column consists of the posterior one-third of the vertebral body and the posterior longitudinal ligament. The posterior column consists of all structures posterior to the posterior longitudinal ligament.

AO Classification System

The AO classification introduced by Magerl[13] in 1994 is based on the mechanism of injury. The following 3 main classifications exist: Type A–compression injury, Type B–distraction injury, and Type C–translational/rotational injury. The disadvantage of the AO classification is that it is extremely complex, including more than 53 different mechanisms of injury, thus limiting its usefulness in the clinical setting. Despite their shortcomings, all 3 systems attempt to describe the pattern of injury based on anatomy or injury mechanism, and attempt to aid clinicians in diagnosis and treatment. There is moderate inter- and intraobserver agreement among these 3 classification systems.[14]

Thoracolumbar Injury Classification and Severity Score

TLICS was recently described by the Spine Trauma Study group as a comprehensive and practical classification system that is based on the injury morphology, status of the posterior ligamentous complex, and neurologic status of the patient.[1] A severity score is determined in conjunction

Figure 13-2. A lateral lumbar plain radiograph of a 76-year-old woman, who fell down 3 stairs, demonstrates compression fractures of T12, L2, and L4 (solid white arrows). Note that the compression fractures affect the anterior column, with significant loss of height of the anterior portion of the involved vertebral bodies.

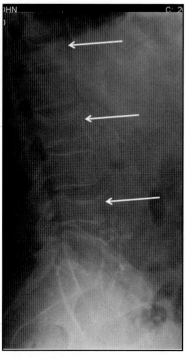

with the TLICS classification system to guide treatment decisions. Injury morphology is divided into compression, translation/rotation, and distraction injuries.

Compression injuries are the result of the vertebral body failing under axial loading and typically involve the anterior spinal column (Figure 13-2). Compression injuries with a burst component (ie, involving the anterior and middle columns) are more significant and are therefore assigned an additional point when determining the severity score (Figure 13-3). An additional point is also assigned if there is a lateral flexion component with a resultant coronal plane deformity of 15 degrees or greater.[1] In summary, a compression fracture would be assigned 1 point; a burst fracture, 2 points; and a compression fracture with a coronal plane deformity of 15 degrees, 2 points[1,15-19] (Table 13-3).

Translational or rotational injuries are generally the result of torsional and shear forces acting on the spinal column. These injuries represent a more significant instability pattern

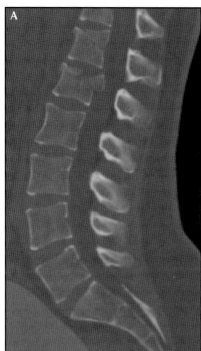

Figure 13-3. (A) Sagittal and (B) axial CT images of a 22-year-old woman who fell out of a second story window, demonstrate an L1 burst fracture. Note that there is compromise of both the anterior and middle columns of the spine, with compression of the vertebral body and significant posterior retropulsion of bone into the spinal canal. The axial image shows the retropulsed bone causes approximately 40% spinal canal compromise. The mechanism of injury in this type of fracture usually involves a significant axial load to the spine.

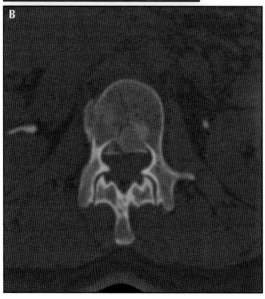

Table 13-3

INJURY MORPHOLOGY

Type	Qualifiers	Points
Compression		+1
	Burst	+1
Translational/rotational		+3
Distraction		+4

Adapted from Vaccaro AR, Lehman RA Jr, Hurlbert RJ, et al. A new classification of thoracolumbar injuries: the importance of injury morphology, the integrity of the posterior ligamentous complex, and neurologic status. *Spine (Phila Pa 1976).* 2005;30(20):2325-2233.

than injuries resulting from a pure compression mechanism. Injuries sustained via a translational or rotational mechanism are assigned 3 points (see Table 13-3). The distinguishing feature of a distraction injury is that the spine fails under tensile stresses (Figure 13-4). The failure of the spinal column can be purely osseous or ligamentous in nature, or a combination of both. These injuries are unstable owing to circumferential disruption of the spinal elements. This injury is assigned 4 points (see Table 13-3).[1]

The posterior ligamentous complex of the thoracolumbar spine consists of the following soft-tissue stabilizing structures: facet capsules, ligamentum flavum, interspinous ligament, supraspinous ligament, and thoracodorsal fascia. Injury to these structures has been shown to compromise the stability of the spine. Within the TLICS classification system, the integrity of the posterior ligamentous complex is categorized as intact, indeterminate disruption, or definitely disrupted.[1] The posterior capsular ligamentous complex historically has been assessed by clinical examination (palpable gap between spinous processes), interspinous widening on plain films, or reconstructed CT evaluation. A recent study has shown MRI and CT to be useful in evaluating posterior ligamentous complex status in the setting of normal appearing radiographs.[20] MRI assessment in the presence of posterior ligamentous disruption may demonstrate hyperintensity in the posterior ligament complex on fat-suppressed T2-weighted images (Figure 13-5). The absence

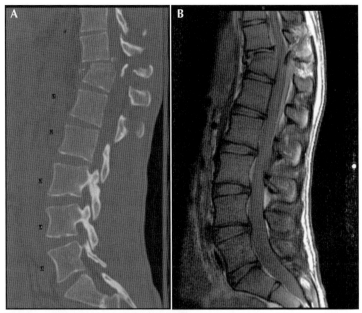

Figures 13-4. (A) Sagittal CT and (B) sagittal MRI of a 26-year-old man who was involved in a high-speed, head-on motor vehicle collision. At the time of presentation, he was assigned an AIS grade of "B." These images demonstrate a flexion-distraction injury at the T11-T12 level, with anterior translation of T11 on T12, disruption of the T11-T12 intervertebral disk, fracture-dislocation of the T11-T12 facet joints, and fractures through the lamina and spinous processes of T11 and T12. Significant spinal cord compression is noted on MRI (B).

of injury to the posterior ligamentous complex is assigned 0 points, an indeterminate disruption is assigned 2 points, and a definite disruption of the posterior ligamentous complex is assigned 3 points (Table 13-4).

The neurologic status of the patient is divided into 4 subcategories based on the severity of the deficit as well as the potential for neural recovery with surgical intervention: intact, nerve root injury, complete spinal cord injury (motor and sensory), and incomplete spinal cord injury (motor or sensory) or cauda equina syndrome.[1] Patients with an intact neurologic examination are assigned 0 points, patients with a nerve root injury and patients with a complete spinal cord injury are assigned

Figure 13-5. Sagittal MRI demonstrating intact posterior ligamentous complex components (ie, ligamentum flavum, interspinous ligament, and supraspinous ligament) at the L1-L2 level (solid white circle) and complete disruption of these same posterior ligamentous complex components at the T11-T12 level. This injury was the result of a flexion-distraction mechanism.

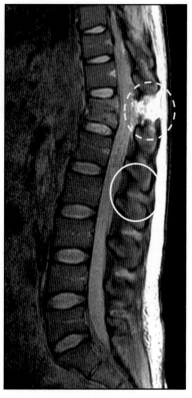

2 points. Patients with an incomplete spinal cord injury or cauda equina syndrome are assigned 3 points (Table 13-5).

To determine the total TLICS score, the subscore from each of the 3 primary axes (ie, morphology, posterior ligamentous complex status, and neurologic status) are added together. The total score can be helpful in guiding treatment of thoracolumbar injuries.

TREATMENT OF THORACOLUMBAR INJURY

High-Dose Methylprednisolone

The use of high-dose methylprednisolone in the setting of acute spinal cord injury is controversial. Several publications

Table 13-4

INTEGRITY OF THE POSTERIOR LIGAMENTOUS COMPLEX

Posterior Ligamentous Complex Disrupted in Tension, Rotation, or Translation	Points
Intact	0
Suspected/indeterminate	2
Injured	3

Adapted from Vaccaro AR, Lehman RA Jr, Hurlbert RJ, et al. A new classification of thoracolumbar injuries: the importance of injury morphology, the integrity of the posterior ligamentous complex, and neurologic status. *Spine (Phila Pa 1976).* 2005;30(20):2325-2233.

Table 13-5

NEUROLOGIC STATUS

Involvement	Qualifiers	Points
Intact		0
Nerve root		2
Cord, conus medullaris	Complete	2
	Incomplete	3
Cauda equina		3

Adapted from Vaccaro AR, Lehman RA Jr, Hurlbert RJ, et al. A new classification of thoracolumbar injuries: the importance of injury morphology, the integrity of the posterior ligamentous complex, and neurologic status. *Spine (Phila Pa 1976).* 2005;30(20):2325-2233.

resulted from the National Acute Spinal Cord Injury Study that have supported the use of methylprednisolone in this setting.[21-23] In patients who present within 3 hours from injury, the recommended dose of methylprednisolone is 30 mg/kg administered intravenously over 15 minutes, followed by

5.4 mg/kg/hour for 23 hours. The steroid infusion should be continued for a total of 48 hours in patients who present between 3 and 8 hours from injury.[21-23] Steroids are not indicated in patients with acute spinal cord injury who present greater than 8 hours from injury. The use of steroids in the setting of acute spinal cord injury remains controversial, and numerous studies suggest that the use of steroids to treat the acute spinal cord injury patient provides no added benefit and may actually be detrimental to the patient.[24-26]

Operative Versus Nonoperative Treatment

The majority of thoracolumbar injuries can be managed conservatively. Most compression and burst fractures are amenable to treatment in a thoracolumbosacral orthosis for a period of 8 to 12 weeks. Fractures that are unstable or are associated with a neurologic deficit often require surgery. The TLICS score can be used to help guide treatment and determine between nonoperative and operative treatment as well as determine which surgical approach (anterior, posterior, or both) is required. Surgical treatment is indicated in patients with a total TLICS score of ≥5 and nonsurgical treatment is indicated in patients with a total TLICS score of ≤3. Surgical or nonsurgical treatment may be appropriate in patients with a total TLICS score of 4.[1] In these cases, the preference of the treating physician, the preference of the patient, and the characteristics of the patient, including medical comorbidities, can help further guide treatment.

The TLICS system is also useful to determine the appropriate approach to surgery in those patients who require surgical intervention. The integrity of the posterior ligamentous complex and the neurologic status of the patient are the 2 most important factors to consider when determining whether to perform anterior, posterior, or a combined anterior-posterior approach to surgical management.[27] The presence of an incomplete neurologic deficit with compression anteriorly typically requires an anterior approach to decompress the neural elements. Patients with injury to the posterior ligamentous complex typically require posterior stabilization. Those with both a neurologic deficit and a disrupted posterior ligamentous complex will likely require a combined anterior-posterior surgical decompression and fusion (Figure 13-6). The appropriate

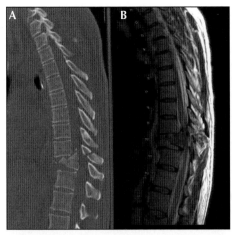

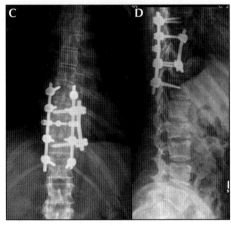

Figure 13-6. (A) Sagittal CT and (B) MRI demonstrating a flexion-istraction injury in a 50-year-old woman who fell off of a 25-foot ladder. This patient was assigned an AIS grade of "A" (ie, complete neurologic injury). The images demonstrate translation of T9 on T10, fracture of the T10 vertebral body and T9 posterior elements. Because this patient had a neurologic deficit, incompetent anterior column, and complete disruption of the posterior ligamentous complex, she was treated with a combined anterior-posterior approach. Postoperative AP. (C) Lateral and (D) plain radiographs show an anterior T10 corpectomy with placement of a structural allograft iliac crest from T9 to T11 and an anterior single rod and screw construct, as well as a posterior spinal fusion from T8 to T12 with pedicle screws and rods, and a crosslink. Note in the images that the patient also has compression fractures of L2 and L3, which were managed nonoperatively in a brace.

Table 13-6

HELPFUL HINTS

- Mechanism of injury is very important to understanding fracture pattern
- Physical examination should include full neurological evaluation
    - Evidence of thoracolumbar fracture/injury on imaging
    - X-ray: Interspinous process widening, vertebral body translation/angulation, splaying of the pedicles
    - CT scan: very detailed image of bony elements, sensitive for detecting fractures
    - MRI: shows neural elements, intervertebral disc, and spinal ligaments in great detail, very sensitive for detecting soft tissue injury and for demonstrating compression of the neural elements
- Accurate classification of injury will guide treatment decisions and is helpful in surgical planning
- TLICS classification is the most recently described classification system
- Treatment
    - Most injuries treatment in brace
    - Surgical treatment indicated for
    - Neurological injury
    - Unstable spine injury

timing of surgical decompression (ie, early versus late) in a patient with neurologic deficit remains a matter of debate.

CONCLUSION

Thoracolumbar injuries are usually the result of high-energy trauma. It is important to ensure that thoracolumbar injury patients are stable before a thorough neurologic and spinal examination is performed. It is also important to diagnose and treat concomitant injuries (eg, bowel, liver, and splenic injury), which are fairly common in this patient population. A thorough evaluation includes a history, physical examination, and imaging studies (ie, radiographs, CT, and MRI; Table 13-6).

Thoracolumbar injuries are classified according to the morphology of the injury, the status of the posterior ligamentous complex, and the neurologic status of the patient. Although the majority of thoracolumbar injuries can be treated nonoperatively with a brace, injuries that cause significant spinal instability and neurologic compromise will often require surgical management.

REFERENCES

1. Vaccaro AR, Lehman RA Jr, Hurlbert RJ, et al. A new classification of thoracolumbar injuries: the importance of injury morphology, the integrity of the posterior ligamentous complex, and neurologic status. *Spine (Phila Pa 1976)*. 2005;30(20):2325-2233.
2. Priebe MM, Waring WP. The interobserver reliability of the revised American Spinal Injury Association standards for neurological classification of spinal injury patients. *Am J Phys Med Rehabil*. 1991;70(5): 268-270.
3. Marino RJ, Barros T, Biering-Sorenson F, et al. International standards for neurological classification of spinal cord injury. *J Spinal Cord Med*. 2003;(26 suppl 1):S50-S56.
4. Savitsky E, Votey S. Emergency department approach to acute thoracolumbar spine injury. *J Emerg Med*. 1997;15(1):49-60.
5. Delamarter RB, Coyle J. Acute management of spinal cord injury. *J Am Acad Orthop Surg*. 1999;7(3):166-175.
6. Ditunno JF, Little JW, Tessler A, Burns AS. Spinal shock revisited: a four-phase model. *Spinal Cord*. 2004;42(7):383-395.
7. Samuels LE, Kerstein MD. 'Routine' radiologic evaluation of the thoracolumbar spine in blunt trauma patients: a reappraisal. *J Trauma*. 1993;34(1):85-89.
8. Lee HM, Kim HS, Kim DJ, Suk KS, Park JO, Kim NH. Reliability of magnetic resonance imaging in detecting posterior ligament complex injury in thoracolumbar spinal fractures. *Spine (Phila Pa 1976)*. 2000;25(16): 2079-2084.
9. Vaccaro AR, Rihn JA, Saravanja D, et al. Injury of the posterior ligamentous complex of the thoracolumbar spine: a prospective evaluation of the diagnostic accuracy of magnetic resonance imaging. *Spine (Phila Pa 1976)*. 2009;34(23):E841-E847.
10. Holdsworth F. Fractures, dislocations, and fracture-dislocations of the spine. *J Bone Joint Surg Am*. 1970;52(8):1534-1551.
11. Denis F. The three column spine and its significance in the classification of acute thoracolumbar spinal injuries. *Spine (Phila Pa 1976)*. 1983;8(8):817-831.
12. Ferguson RL, Allen BL Jr. A mechanistic classification of thoracolumbar spine fractures. *Clin Orthop Relat Res*. 1984;(189):77-88.
13. Magerl F, Aebi M, Gertzbein SD, Harms J, Nazarian S. A comprehensive classification of thoracic and lumbar injuries. *Eur Spine J*. 1994;3(4): 184-201.

14. Oner FC, Ramos LM, Simmermacher RK, et al. Classification of thoracic and lumbar spine fractures: problems of reproducibility: a study of 53 patients using CT and MRI. *Eur Spine J*. 2002;11(3):235-245.

15. Vaccaro AR, Zeiller SC, Hulbert RJ, et al. The thoracolumbar injury severity score: a proposed treatment algorithm. *J Spinal Disord Tech*. 2005;18(3):209-215.

16. Vaccaro AR, Lehman RA Jr, Hurlbert RJ, et al. A new classification of thoracolumbar injuries: the importance of injury morphology, the integrity of the posterior ligamentous complex, and neurologic status. *Spine (Phila Pa 1976)*. 2005;30(20):2325-2333.

17. Wood KB, Khanna G, Vaccaro AR, Arnold PM, Harris MB, Mehbod AA. Assessment of two thoracolumbar fracture classification systems as used by multiple surgeons. *J Bone Joint Surg Am*. 2005;87(7):1423-1429.

18. Vaccaro AR, Baron EM, Sanfilippo J, et al. Reliability of a novel classification system for thoracolumbar injuries: the Thoracolumbar Injury Severity Score. *Spine (Phila Pa 1976)*. 2006;31(11)(suppl):S62-S69.

19. Bono CM, Vaccaro AR, Hurlbert RJ, et al. Validating a newly proposed classification system for thoracolumbar spine trauma: looking to the future of the Thoracolumbar Injury Classification and Severity Score. *J Orthop Trauma*. 2006;20(8):567-572.

20. Lee JY, Vaccaro RJ, Schweitzer KM Jr, et al. Assessment of injury to the thoracolumbar posterior ligamentous complex in the setting of normal-appearing plain radiography. *Spine J*. 2007;7(4):422-427.

21. Bracken MB. Methylprednisolone in the management of acute spinal cord injuries. *Med J Aust*. 1990;153(6):368.

22. Bracken MB, Shepard MJ, Hellenbrand KG, et al. Methylprednisolone and neurological function 1 year after spinal cord injury: results of the National Acute Spinal Cord Injury Study. *J Neurosurg*. 1985;63(5):704-713.

23. Bracken MB, Shepard MJ, Holford TR, et al. Administration of methylprednisolone for 24 or 48 hours or tirilazad mesylate for 48 hours in the treatment of acute spinal cord injury: results of the Third National Acute Spinal Cord Injury Randomized Controlled Trial. National Acute Spinal Cord Injury Study. *JAMA*. 1997;277(20):1597-1604.

24. Hurlbert RJ. Methylprednisolone for acute spinal cord injury: an inappropriate standard of care. *J Neurosurg*. 2000;93(1)(suppl):1-7.

25. Hurlbert RJ. The role of steroids in acute spinal cord injury: an evidence-based analysis. *Spine (Phila Pa 1976)*. 2001;26(24)(suppl):S39-S46.

26. Hurlbert RJ. Strategies of medical intervention in the management of acute spinal cord injury. *Spine (Phila Pa 1976)*. 2006;31(11)(suppl):S16-S21.

27. Vaccaro AR, Lim MR, Hurlbert RJ, et al. Surgical decision making for unstable thoracolumbar spine injuries: results of a consensus panel review by the Spine Trauma Study Group. *J Spinal Disord Tech*. 2006;19(1):1-10.

14

ADOLESCENT
IDIOPATHIC SCOLIOSIS

Joseph P. Gjolaj, MD and W. Timothy Ward, MD

INTRODUCTION

Scoliosis is defined as an abnormal lateral (coronal plane) curvature of the spine >10 degrees. It can be classified based on etiology, which includes congenital, neuromuscular, syndrome-related, or idiopathic. Most scoliosis, up to 80%, is idiopathic. Although this condition has been extensively researched, its etiology remains unknown. Multiple theories have been proposed including genetic factors, growth abnormalities, and central nervous system causes, as well as disorders of the bone, muscle, and disk. A polygenetic etiology with variable expression is the most likely explanation.

Rihn JA, Harris EB. *Musculoskeletal Examination of the Spine: Making the Complex Simple* (pp. 242-260).
© 2011 SLACK Incorporated.

Table 14-1

HELPFUL HINTS: CLASSIFICATION OF IDIOPATHIC SCOLIOSIS

Type	Age Group	Characteristics
Infantile	0 to 3 years	Rib vertebral angle difference (RVAD) predicts progression
Juvenile	4 to 10 years	Boys affected earlier than girls (5 years versus 7 years)
Adolescent	>10 years	Prevalence up to 3%, curves <40 degrees amenable to bracing

IMAGING

Image	Pertinent Image Views	Findings
Plain radiograph	Full-length (3 foot film), standing posteroanterior, and lateral	Scoliosis (thoracic, thoracolumbar, and lumbar); kyphosis
Magnetic resonance imaging	Sagittal, coronal, and axial T1-weighted and T2-weighted	Syrinx; meningocele
Computed tomography	Axial, coronal, and sagittal reformats and 3D reconstruction	Bony anatomy; pedicle size; rotation

Idiopathic scoliosis can be further classified by age at diagnosis. Infantile scoliosis is diagnosed between 0 and 3 years; juvenile, between 4 and 10 years; and adolescent, >10 years (Table 14-1). Adolescent idiopathic scoliosis is the most frequently encountered form of scoliosis (up to 80%) and will be the focus of this chapter.

Scoliotic curves may be classified as major or minor. The major curve is the curve of the greatest magnitude and is considered the first to develop. Minor or "compensatory" curves develop after the primary curve and function to balance the head and trunk over the pelvis. Curves may also be described as structural or nonstructural, with structural curves being more rigid (not easily correctable to < 25 to 30 degrees with lateral bending).

HISTORY

Evaluating a patient with scoliosis requires knowledge of all of the conditions that are associated with scoliosis and their particular history and physical examination findings. Although scoliosis in most adolescents will be idiopathic in nature, a thorough history and physical examination are crucial to rule out other causes.

As a part of history taking, the physician should include questions about the patient's recent growth, physical signs of puberty (ie, axillary and pubic hair, breast budding, onset of menses in girls, and voice changes in boys), as well as family history of scoliosis. Idiopathic scoliosis occurs 3 times more frequently in children whose parents have scoliosis and 7 times more frequently if a sibling has scoliosis.[1] A record of height increase during the past few years is helpful to predict remaining spinal growth as well as risk for progression of scoliosis.[2] A history of prior surgeries (ie, prior thoracotomy) may be useful in identifying scoliosis associated with congenital heart disease. Also, a thorough review of systems can identify disorders associated with scoliosis.

Other important findings in the patient history include severe back pain, presence of neurologic symptoms, and age of onset of scoliosis, as well as rate of curve progression. An adolescent with severe back pain and scoliosis, particularly if accompanied by myelopathic or radiculopathic symptoms, not only requires a detailed history and physical examination but also radiographic studies such as magnetic resonance imaging (MRI), computed tomography, or computed tomography myelography since an underlying etiology may be found in these cases.[3,4]

The presence of neurologic symptoms can be caused by intraspinal abnormalities such as syringomyelia with or without Chiari malformation, tethered spinal cord, diastematomyelia, or intraspinal tumors. Radicular symptoms including leg or arm pain, numbness, or weakness, as well as changes in bowel or bladder function, warrant further imaging investigation. Neurologic history should include questions about patient's difficulties with grasping, writing, walking, and climbing stairs. In patients < 10 years, spinal curvature may be the presenting sign of a neural axis abnormality.[5] Finally, the rapid development of a severe curvature may indicate a nonidiopathic etiology.

EXAMINATION

Physical examination specific to scoliosis includes not only evaluation of the patient's trunk and spine but also attention directed to aspects of the neurologic system, skin, and extremities, particularly potential limb-length inequality (Table 14-2).

Spinal Examination

The spinal examination should include an assessment of truncal shape and balance. The adolescent should be clothed in either a halter top or a gown covering the anterior torso with the arms and entire back uncovered.

Inspection of the posterior torso in the erect standing position is directed toward the findings of shoulder, trapezial or scapular asymmetry, and thorax or flank asymmetry. A plumb line dropped from the C7 spinous process should fall in line with the gluteal cleft. Any translation of the plumb line off of the midline indicates a component of truncal imbalance.

The lateral aspect of the right and left rib cage should be symmetrically aligned with the ipsilateral iliac crest regions. Asymmetry of these landmarks may indicate the presence of a thoracic scoliosis. Even a small lumbar curve may fill out the ipsilateral flank, resulting in marked asymmetry with prominence of the contralateral iliac crest. Patients often misinterpret this visual finding and express concern that they believe their hips to be asymmetrical.

With the adolescent facing the examiner, the gown can be elevated to a level just under the breasts. This view gives a good look at any lumbar flank asymmetry which, if present, is often the major complaint for the patient with lumbar scoliosis. Anterior inspection will highlight lumbar flank asymmetry more dramatically than from the posterior view.

It is not generally necessary to observe the breasts or genital area in girls as part of a routine scoliosis examination. However, it is helpful to ask the adolescent girl if she has any perception that her breasts are asymmetrical in size. In scoliosis, the spine and rib cage rotate in the axial plane so that the breast on the concave side of the measured Cobb angle lies in a more anterior position compared to the

Table 14-2

METHODS FOR EXAMINING: TESTS AND FINDINGS

Test

Shoulder and
scapular assymetry

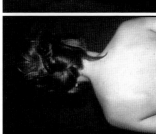

Description

Examining the standing patient
from the front and back may
reveal shoulder height asymmetry
and/or scapular asymmetry, indicating the presence of scoliosis.

(continued)

Table 14-2 (continued)

METHODS FOR EXAMINING: TESTS AND FINDINGS

Test		Description
Waist and flank asymmetry		Asymmetric appearance of the waist or flank in the standing patient is also seen with scoliosis.

(continued)

Table 14-2 (continued)

METHODS FOR EXAMINING: TESTS AND FINDINGS

Test		Description
Plumb line	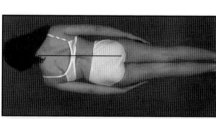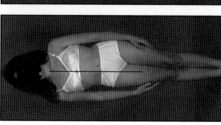	A "plumb" or gravity line dropped from the C7 prominence should fall in line with the gluteal cleft posteriorly. Translation of the plumb line off of the midline indicates the presence of truncal imbalance.

(continued)

Table 14-2 (continued)

METHODS FOR EXAMINING: TESTS AND FINDINGS

Test		Description
Adams forward bending test	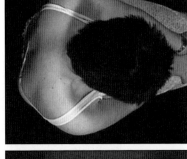	The patient bends forward at the waist, with the knees straight and palms together. Examination from behind assesses lumbar and mid-thoracic rotation, from the front assesses upper thoracic rotation, and from the side assesses kyphosis.

(continued)

Table 14-2 (continued)

METHODS FOR EXAMINING: TESTS AND FINDINGS

Test		Description
Paralumbar prominence	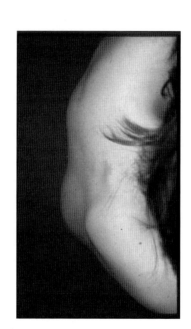	Also identified during the forward bending test, paralumbar prominence can indicate significant rotational deformity associated with lumbar scoliosis.

contralateral breast. This axial malrotation is interpreted by the patient as having a larger breast on one side.

Adams[6] first described the forward bending test in 1865, and it is still in use today. In this test, the patient bends forward at the waist, with the knees straight and palms together. The examiner evaluates the patient from behind to assess lumbar and midthoracic rotation, from the front to assess upper thoracic rotation, and from the side to assess kyphosis. To quantify the deformity, a scoliometer may be used to measure the angle of trunk rotation.[7] Generally, a trunk rotation angle of 5 to 7 degrees is associated with a Cobb angle of 15 to 20 degrees.[8] The height of the prominence in centimeters can also quantify the rotational deformity. In large cases of scoliosis, the parathoracic or lumbar prominences can be dramatic.

In the Adams position, the spinous processes can also be palpated, and any deviation from a straight line also is an indication of vertebral malrotation. The Adams forward bending test is used by school personnel to screen for scoliosis. It is possible for patients without scoliosis to have small amounts of parathoracic or paralumbar asymmetry as well as shoulder or waist asymmetry without the presence of a Cobb measurement scoliosis by radiograph. The subtle findings in this setting simply reflect normal body asymmetry that can be seen in up to 40% of the population.

Limb-length inequality contributing to apparent scoliosis can be assessed by placing the examiner's hands on the midlateral iliac crests or posterior superior iliac spines to assess pelvic tilt due to the inequality. Any inequality should be quantified with the use of measured blocks under the short leg. Observance of the level of the right and left sacral dimples will also give an indication of any limb-length inequality. Finally, in patients who have a pelvic tilt, limb lengths should be measured while the patient is supine. This may help identify a compensatory spinal curvature due to limb-length inequality.

A thorough scoliosis examination should also include inspection of the skin for abnormalities such as café-au-lait spots or axillary freckling, which can be associated with neurofibromatosis. Dimpling of the skin or a hairy patch in the lumbosacral region may indicate spina bifida. In addition, skin or joints that have excessive laxity may be related to connective tissue disorders such as Marfan syndrome or Ehlers-Danlos syndrome.

Neurologic Examination

During the neurologic examination, the patient's balance, sensation, and motor strength in all extremities should be evaluated. Balance may be specifically tested by watching the patient's gait, heel-and-toe walking, and single-leg hop. Deep tendon reflexes should be tested in both upper and lower extremities. Abnormalities in Babinski's reflex and abdominal reflexes may uncover intraspinal disorders such as syringomyelia.

PATHOANATOMY

The normal spine, although straight in the coronal plane, does have curvatures in the sagittal plane. Specifically, the normal spine has thoracic kyphosis averaging 30 to 35 degrees and lumbar lordosis averaging 50 to 60 degrees.[9] The scoliotic spine differs from the normal spine not only in its coronal plane curvature but also in its vertebral rotation at the apex of the curve. This rotation in the thoracic spine produces a typical chest wall prominence (Adams sign) often seen in scoliosis.

Previously, it was thought that scoliosis was also associated with kyphosis, which explained the characteristic "hump" seen in these patients. This hump or rib prominence is caused by rotational deformity of the rib cage and vertebrae. In fact, most thoracic idiopathic scoliosis is associated with decreased thoracic kyphosis.[10] The etiology of this loss of kyphosis remains unknown. Incidentally, most progressive idiopathic scoliosis is also convex to the right side.[11] Etiology of this finding is also unknown.

Scoliosis also displays local deformity at the vertebral and disk levels. Wedging is noted in both structures, and subsequent changes to the vertebral bodies occur consistent with normal bone biomechanics. Specifically, reduced growth is observed in areas of excessive compression as seen in the concavity of the scoliotic spine. These factors contribute to asymmetric growth as well as remodeling of all elements of the spine, including the vertebral bodies, laminae, pedicles, facet joints, and transverse and spinous processes. This reduced concave growth exacerbates the existing deformity, increasing the compressive forces, which in turn perpetuates the entire process.

Figure 14-1. The Cobb method for measurement of coronal plane deformity. The end vertebrae of the curve being measured are identified. The end vertebrae are defined as the vertebrae at the upper and lower limits of the curve that tilt most severely toward the concavity of the curve. Lines are then drawn parallel to the superior end plate of the upper end vertebra and the inferior end plate of the lower end vertebra (solid black lines). Lines are then drawn perpendicular to the end-plate lines (broken black lines). The angle formed by theses lines (black arrow) represents the Cobb angle.

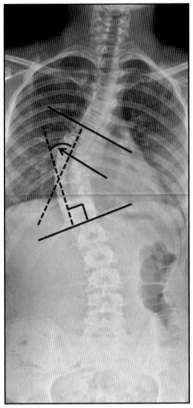

IMAGING

When screening for scoliosis, the initial radiographic evaluation should always include a standing posteroanterior (PA) radiograph of the entire spine down to the pelvis on a single cassette. Typically, this requires a 3-foot length film. The coronal plane curvature is measured on the standing PA radiograph according to the Cobb method, which is described in detail in Figure 14-1. The standing PA radiograph may also help estimate skeletal maturity. Risser[12] has described the most widely used method, which uses the iliac apophysis as a guide. A lateral film is not required at initial screening of a thoracic curve, unless sagittal deformity or back pain is present. A lateral radiograph of the lumbosacral junction is also recommended in lumbar scoliosis to evaluate for spondylolysis or spondylolisthesis.

When surgical treatment is appropriate, lateral bending radiographs as well as a standing lateral radiographs are required. The lateral bending radiographs help assess curve flexibility and therefore assist in operative planning. There are a number of ways a lateral bending radiograph may be obtained. Supine-position lateral bending radiographs are a commonly used technique. Others prefer the standing-position lateral bending radiographs, advocating that this technique is a better indicator of correction in the lumbar spine. Still others recommend lateral bending radiographs taken over a bolster, believing the greater correction seen on these films better corresponds to the correction obtainable with modern surgical instrumentation.[13] In larger curves (>60 to 70 degrees), longitudinal traction films may be beneficial.[14] Currently, there is no standard method for obtaining lateral bending radiographs. There is also little consensus on how to interpret the information obtained from these studies.

A number of prospective studies have been published assessing routine MRI screening for preoperative evaluation of idiopathic scoliosis patients.[15] So far, there has been no evidence that an MRI is useful in typical adolescent idiopathic scoliosis. However, in patients with neurologic abnormalities or cutaneous skin findings, an MRI is recommended.[16] In addition, hyperkyphotic thoracic alignment may be related to syringomyelia that can be identified by MRI.[17] Patients with significant back pain with no obvious cause may warrant an MRI or bone scan to evaluate for presence of tumor, infection, or spondylolysis.

TREATMENT

Nonoperative Treatment

Because most adolescent idiopathic scoliosis patients have curves <20 degrees that do not progress, monitoring is all that is indicated. Curves <25 degrees should be monitored every 6 to 12 months with serial radiographs and repeat clinical examination. Patients in the rapid phases of growth should be seen more frequently (every 4 to 6 months). For curves >45 to 50 degrees in which surgery has not been performed, monitoring should continue after skeletal maturity, with serial radiographs every 5 years.

Orthotic (brace) treatment is used by most pediatric ortho-pedists in the growing child with a progressive curve of 25 to 30 degrees.[18] If enough spinal growth remains, some physi-cians will even recommend brace treatment if a curve reaches 20 degrees. The effectiveness of brace treatment depends on spinal growth, thus bracing is typically recommended for patients with substantial growth remaining. Multiple retrospective, low-evidence studies have concluded that brac-ing is effective in controlling curve progression; however, brac-ing does have limitations, and its effectiveness has recently been called into question. Patients with curves >45 degrees are not considered to be candidates for brace treatment because control of progression is generally not seen in a curve of this magnitude. Also, even in studies that conclude bracing is an effective form of treatment, success is defined as the ability to maintain the curve at the degree of severity that was present at the initiation of bracing.[19] After bracing is discontinued, the curve typically returns to the pretreatment level of severity.

A number of different braces have been developed. The Milwaukee brace is one of the earliest braces developed. Although not cosmetically favorable (because of its neckpiece and uprights), the Milwaukee brace is one of the few braces that has any potential to maintain upper thoracic curves.[20] This brace is rarely used in current practice because of its obtrusiveness and the belief that it may be no more effective than other less obtrusive braces for most cases of scoliosis. Underarm braces such as the Boston and Wilmington have become more popular because they are less conspicuous and equally effective.[21]

Despite this, many patients will still be noncompliant with brace wear. Common complaints are pain, discomfort from heat, poor fit, and self-esteem issues. These particular braces are recommended for full-time (23 hours per day) wear. The Charleston nighttime bending brace may have better compli-ance since it is only worn at night.[22] It attempts to correct curves by producing an exaggerated correction that is so extreme it makes daily activities difficult. For this reason, it is only prescribed for nighttime wear.

Operative Treatment

An adolescent with major curve magnitude >45 or 50 de-grees is indicated for operative treatment by most pediatric orthopedic surgeons. Factors affecting the decision to operate include clinical deformity, risk of progression, level of skeletal maturity, and pattern of curve. A flexible minor curve, defined as the smaller curve in a double curve pattern that also bends out <25 degrees, is considered to be nonstructural and generally does not warrant fusion. Conversely, a minor curve that measures >25 to 30 degrees on side bending films is considered structural and is frequently included in the fusion. Severe, rigid curves (not correctable beyond 50 to 60 degrees on lateral bending radiographs may require anterior release procedures prior to definitive posterior instrumentation and fusion.

Multiple methods of surgical correction exist. Posterior spinal instrumentation was introduced by Harrington in the 1960s and proved to be more predictable in obtaining Cobb angle correction and fusion than earlier reports on in situ fusion with body casting (Figure 14-2A).[23] Drawbacks to Harrington rod instrumentation included the continued need for a postoperative cast or brace, risk for development of decreased thoracic kyphosis and flat-back deformity, and inability to apply 3-dimensional corrective forces to the spine. Modern instrumentation methods include the use of multisegmental hooks, wires, and pedicle screws in various configuration patterns. All pedicle screw constructs appear to achieve the greatest degree of Cobb angle correction, but this benefit has not been shown to improved patient reported outcomes over hook or hybrid (hooks, wires, and screw) constructs (Figure 14-2B-G).

Younger patients treated by isolated posterior instrumentation and fusion may be at risk for postfusion crankshaft deformity (continued anterior spinal growth after successful posterior fusion). Furthermore, large curves that measure >70 degrees may be relatively rigid and resistant to correction. In these settings, combined anterior and posterior fusion has been recommended by some surgeons. The anterior procedure involves disk excision and release of the anterior longitudinal ligament. This increases the flexibility of the curve, allows for greater curve correction, and allows for a fusion of the anterior verterbral column, thus minimizing the risk of developing a crankshaft phenomenon.

Complications of operative treatment include neurologic injury, blood loss, implant failure, and loss of correction over time. Although the incidence of neurologic injury after surgery for idiopathic scoliosis is not well documented, in one

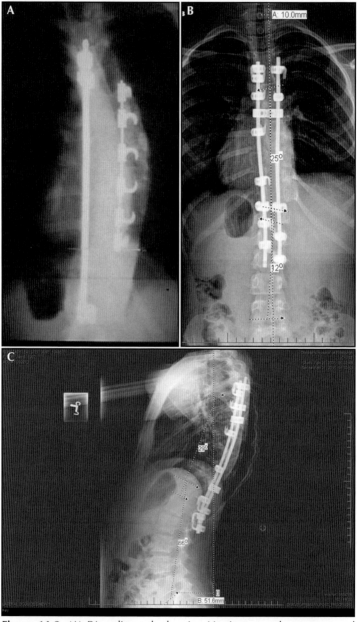

Figure 14-2. (A) PA radiograph showing Harrington rod construct and (B) PA radiograph showing hook construct. (C) Lateral radiographs showing hook construct. *(continued)*

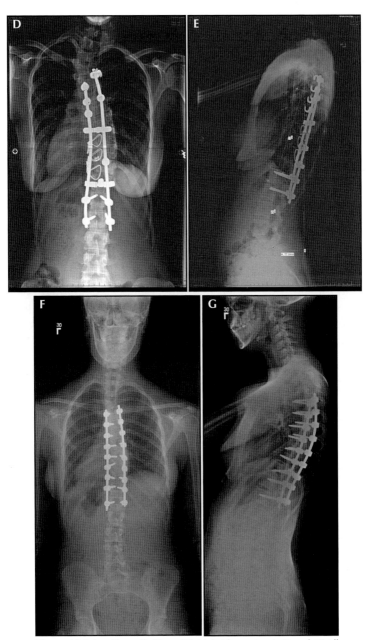

Figure 14-2 (continued). (D and E) Hybrid construct. (F and G) All pedicle screw construct.

large series, the reported incidence was approximately 1% in all types of scoliosis surgery for all ages.[24] For adolescent idiopathic scoliosis surgery, the incidence is probably much less than this. The etiology can be related to either direct trauma to the spinal cord, excessive traction, or vascular insufficiency to the cord. Spinal cord monitoring is now routinely used in efforts to identify neurologic injury intraoperatively and reverse the causes if possible.

CONCLUSION

Adolescent idiopathic scoliosis is a diagnosis of exclusion, and patients can be identified by a careful history and physical examination. Less severe curves are amenable to bracing, which may halt progression of scoliosis in the compliant patient. More severe curves may be successfully corrected with spinal fusion using modern segmental instrumentation methods that obviate the need for postoperative casts or braces. Complications of surgical intervention are uncommon, and outcomes can be excellent in the hands of experienced surgeons.

REFERENCES

1. Wynne-Davies R. Familial (idiopathic) scoliosis: a family survey. *J Bone Joint Surg Br.* 1968;50(1):24-30.
2. Little DG, Song KM, Katz D, Herring JA. Relationship of peak height velocity to other maturity indicators in idiopathic scoliosis in girls. *J Bone Joint Surg Am.* 2000;82(5):685-693.
3. Ramirez N, Johnson CE II, Brown RH. The prevalence of back pain in children who have idiopathic scoliosis. *J Bone Joint Surg Am.* 1997;79(3):364-368.
4. Mehta MH. Pain provoked scoliosis: observations on the evolution of the deformity. *Clin Orthop.* 1978;(135):58-65.
5. Gupta P, Lenke LG, Bridwell KH. Incidence of neural axis abnormalities in infantile and juvenile patients with spinal deformity: is a magnetic resonance image screening necessary? *Spine (Phila Pa 1976).* 1998;23(2):206-210.
6. Adams W. *Lectures on Pathology and Treatment of Lateral and Other Forms of Curvature of the Spine.* London, United Kingdom: Churchill Livingstone; 1865.
7. Bunnell WP. Outcome of spinal screening. *Spine (Phila Pa 1976).* 1993;18(12):1572-1580.

8. Korovessis PG, Stamatakis MV. Prediction of scoliotic Cobb angle with the use of the scoliometer. *Spine (Phila Pa 1976).* 1996;21(14):1661-1666.

9. Bernhardt M, Bridwell KH. Segmental analysis of the sagittal plane alignment of the normal thoracic and lumbar spines and the thoracolumbar junction. *Spine (Phila Pa 1976).* 1989;14(7):717-721.

10. Dickson RA, Lawton JO, Archer IA, Butt WP. The pathogenesis of idiopathic scoliosis: biplanar spinal asymmetry. *J Bone Joint Surg Br.* 1984;66(1):8-15.

11. Goldberg CJ, Dowling FE, Fogarty EE. Left thoracic scoliosis configurations: why so different? *Spine (Phila Pa 1976).* 1994;19(12):1385-1389.

12. Risser JC. The iliac apophysis: an invaluable sign in the management of scoliosis. *Clin Orthop.* 1958;(11):111-119.

13. Cheung KM, Luk KD. Prediction of correction of scoliosis with use of the fulcrum bending radiograph. *J Bone Joint Surg Am.* 1997;79(8):1144-1150.

14. Vaughnn JJ, Winter RB, Lonstein JE. Comparison of the use of supine bending and traction radiographs in the selection of fusion area in adolescent idiopathic scoliosis. *Spine (Phila Pa 1976).* 1996;21(21):2469-2473.

15. Do T, Fras C, Burke S, Widmann RF, Rawlins B, Boachie-Adjei O. Clinical value of routine preoperative magnetic resonance imaging in adolescent idiopathic scoliosis: a prospective study of three hundred and twenty-seven patients. *J Bone Joint Surg Am.* 2001;83(4):577-579.

16. Schwend RM, Hennrikus W, Hall JE, Emans JB. Childhood scoliosis: clinical indications for magnetic resonance imaging. *J Bone Joint Surg Am.* 1995;77(1):46-53.

17. Ouellet JA, LaPlaza J, Erickson MA, Birch JG, Burke S, Browne R. Sagittal plane deformity in the thoracic spine: a clue to the presence of syringomyelia as a cause of scoliosis. *Spine (Phila Pa 1976).* 2003;28(18):2147-2151.

18. Nachemson AL, Peterson LE. Effectiveness of treatment with a brace in girls who have adolescent idiopathic scoliosis: a prospective, controlled study based on data from the brace study of the Scoliosis Research Society. *J Bone Joint Surg Am.* 1995;77(6):815-822.

19. Mellencamp DD, Blount WP, Anderson AJ. Milwaukee brace treatment of idiopathic scoliosis: late results. *Clin Orthop.* 1977;(126):47-57.

20. Blount WP, Schmidt AC, Keever ED, Leonard ET. The Milwaukee brace in the treatment of scoliosis. *J Bone Joint Surg Am.* 1958;40-A(3):511-525.

21. Emans JB, Kaelin A, Bancel P, Hall JE, Miller ME. The Boston bracing system for idiopathic scoliosis: follow-up results in 295 patients. *Spine (Phila Pa 1976).* 1986;11(8):792-801.

22. Price CT, Scott DS, Reed FR Jr, Sproul JT, Riccick MF. Nighttime bracing for adolescent idiopathic scoliosis with the Charleston bending brace: long-term follow-up. *J Pediatr Orthop.* 1997;17(6):703-707.

23. Harrington PR. Treatment of scoliosis: correction and internal fixation by spinal instrumentation. *J Bone Joint Surg Am.* 1962;44(4):591-610.

24. Bridwell KH, Lenke LG, Baldus C, Blanke K. Major intraoperative neurologic deficits in pediatric and adult spine deformity patients: incidence and etiology at one institution. *Spine (Phila Pa 1976).* 1998;23(3):324-331.

15

SCHEUERMANN KYPHOSIS

Kathryn H. Hanna, MD and Eric B. Harris, MD

INTRODUCTION

In 1921 a Danish surgeon, Holger Werfel Scheuermann, published an original work that defined kyphosis dorsalis juvenilis as a fixed dorsal kyphosis, a spinal entity very different from "apprentice kyphosis" or "kyphosis muscularis."[1] Kyphosis dorsalis juvenilis, quickly renamed Scheuermann kyphosis, describes a rigid sagittal deformity of the thoracic spine that cannot be actively corrected. Radiographically, the vertebral bodies are wedged anteriorly.

Scheuermann kyphosis is the most common adolescent form of kyphosis. The prevalence worldwide is between 0.4% and 8%.[2,3] A 20,000-patient screening study of Italian

Rihn JA, Harris EB. *Musculoskeletal Examination of the Spine: Making the Complex Simple* (pp. 261-275).
© 2011 SLACK Incorporated.

students[4] found a 1% rate of Scheuermann kyphosis with a female-to-male prevalence of 1.4:1.

There are 3 types of Scheuermann kyphosis. Type I, the classic form, defines a thoracic deformity with the apex between T7 and T9. This form is often associated with a non-structural compensatory lumbar hyperlordosis. Type II is the thoracolumbar form with lower sagittal apex between T10 and T12. Type III is defined as a lumbar form of kyphosis and has been reported in the active young male population.[5]

HISTORY

Often the diagnosis of Scheuermann kyphosis at a young age is mistakenly diagnosed as a postural kyphosis. Taking a thorough history and complete physical examination are paramount in making the correct diagnosis (Table 15-1).

There is a high association of Scheuermann disease with Turner syndrome, cystic fibrosis, and other disease processes such as malnutrition associated with generalized osteoporosis.[5] A complete birth history including a review of genetic tests would identify this subset of patients.

The average patient presents around puberty, and it is a rare patient who demonstrates clinic signs of disease <10 or 11 years.[6] The patient tells a history of a progressive thoracic kyphosis. The patient may or may not notice an increasing compensatory lumbar lordosis. There can be a history of hamstring tightness and contractures because of increased pelvic tilt related to the lumbar lordosis. Patients complain of pain and tenderness at the apex of their curve. Pain is more common in athletes such as gymnasts and wrestlers.

EXAMINATION

The physical examination of a patient with Scheuermann kyphosis begins with the standard physical examination of any patient presenting with spinal deformity complaints. Overall sagittal and coronal alignment of the patient is examined. Patients will have a thoracic kyphosis and often a compensatory lumbar lordosis. The lumbar lordosis is associated with increased pelvic tilt. The increased pelvic tilt leads

Table 15-1

HELPFUL HINTS: DIFFERENCE BETWEEN SCHEUERMANN AND POSTURAL KYPHOSIS

	Scheuermann	Postural
Description	Progressive kyphotic condition secondary to vertebral structural wedging	Benign condition secondary to poor posture
Radiographic findings	3 consecutive wedged vertebrae with 5 degrees of anterior wedging	Less angulated; kyphosis corrects on hyperextension lateral radiograph
Forward bending test	Sagittal kyphosis with area of fixed sharp angulation	Sagittal kyphosis is gently rounded
Body type	Athletic	Asthenic

to hamstring tightness and contractures, which can be seen on hip flexion and examined using a straight-leg test. In extremely tight hamstrings, a patient may present with a pelvic waddle, which presents as a stiff-legged, short-strided gait with the pelvis rotating with each step. One-third of patients have a mild to moderate scoliosis of 10 to 20 degrees.[6] This can be further assessed on the forward bending test where rib prominence can be observed. Specific to Scheuermann kyphosis is the Adams forward bending test, which demonstrates a thoracic hump, sometimes referred to as a gibbous, that is often a sharp 90-degree curve rather than the rounded back seen in postural kyphosis (Figure 15-1). The patient's spinal flexibility should be examined with flexion, extension, side-to-side bending, and rotation tests.

There may be an area of cutaneous skin pigmentation over the apex of the curve, where the skin is constantly rubbed against the back of a chair. It is rare but not impossible for patients to develop radicular or myelopathic symptoms with their kyphosis.[7] A complete neurologic examination should be obtained to include upper- and lower-extremity deep tendon

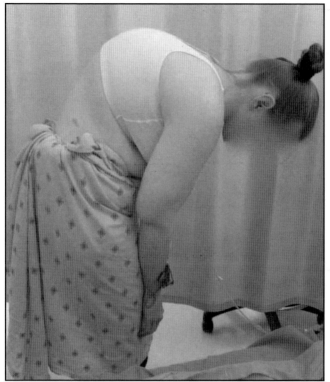

Figure 15-1. Classic Adams forward bending test in a patient with a sharp kyphotic curve.

reflexes, and complete sensory and motor strength examinations, as well as upper motor neuron tests (ie, Hoffmann's and Babinski's signs; Table 15-2).

PATHOANATOMY

There is no definite etiology known for Scheuermann kyphosis. There are many theories ranging from avascular necrosis of the vertebral ring apophysis proposed by Scheuermann, to end-plate deterioration and wedging as a result of herniation of intervertebral material proposed by Schmorl in 1930, to a mechanical theory of wedging caused by repetitive heavy labor and, finally, genetic and endocrine

Table 15-2

Methods for Examining: Patients With Scheuermann Kyphosis

Examination	Technique	Illustration	Grading	Significance
Gait	Patient takes normal strides			Tight hamstrings may lead to a stiff-legged, short-strided, pelvic waddle
Straight-leg raise	Patient is supine; with the contralateral limb resting with neutral hip and extended knee, ipsilateral hip is passively flexed, trying to keep the knee extended			With a positive finding, the patient will be unable to achieve 90 degrees of hip flexion

(continued)

Table 15-2 (continued)

METHODS FOR EXAMINING: PATIENTS WITH SCHEUERMANN KYPHOSIS

Examination	Technique	Illustration	Grading	Significance
Adams forward bending test	With the feet together and knees straight, the patient bends forward and dangles arms	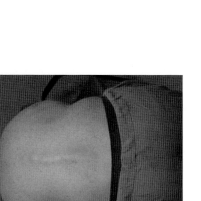		1) Prominent sharp thoracic hump indicates Scheuermann kyphosis 2) Rib prominence or other deformity suggests scoliosis

(continued)

Table 15-2 (continued)

Methods for Examining: Patients With Scheuermann Kyphosis

Examination	Technique	Illustration	Grading	Significance
Range of motion	Flexion, extension, side bending and rotation			Loss of motion may be related to pain or may also represent rigid deformity

(continued)

Table 15-2 (continued)

METHODS FOR EXAMINING: PATIENTS WITH SCHEUERMANN KYPHOSIS

Examination	Technique	Illustration	Grading	Significance
Strength testing	Manual strength testing: upper- and lower-extremity flexion, extension, abduction, and adduction		Grade 5 = full strength; grade 4 = resistance, < full strength; grade 3 = movement against gravity; grade 2 = movement without gravity; grade 1 = visible muscle firing	Loss of strength can represent radiculopathy or myelopathy
Hoffmann's sign	Reflex test; flick terminal phalanx of third or fourth finger	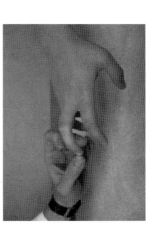	Positive response is flexion of terminal phalanx of thumb; represents injury to corticospinal tract	

(continued)

Table 15-2 (continued)

METHODS FOR EXAMINING: PATIENTS WITH SCHEUERMANN KYPHOSIS

Examination	Technique	Illustration	Grading	Significance
Babinski's sign	Blunt instrument rubbed from heel along lateral plantar foot across metatarsal foot pad	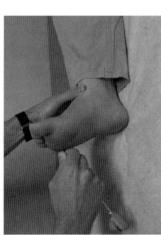		Indicates upper motor neuron finding when toes are "up going," (ie, hallux dorsiflexes and toes span outward)

causes.[8] Twin studies have shown a pair-wise concordance for the disease that is higher between monozygotic twins.[9] There is also evidence that there is a pattern of dominant inheritance between generations.[10]

Histology studies have demonstrated marked end-plate irregularities with Schmorl nodules but have failed to demonstrate evidence of avascular necrosis of the ring apophysis. The end plates of the vertebral bodies have been shown to have abnormal cartilage containing abnormal collagen matrix, which can lead to stunting of the end-plate ossification, allowing for disk material extrusion and possible collapse.[11] On autopsy, the anterior longitudinal ligament has been found to be markedly thickened and bowstrung across the vertebral bodies.[12]

IMAGING

The diagnosis of Scheuermann kyphosis is made on plain radiographs. A typical thoracic series of posteroanterior and lateral views should be obtained (Figure 15-2). The patient should elevate his or her arms at 90 degrees in front of the body to avoid the bony outlines of the upper extremity from obscuring the thoracic spine.[6] In 1964, Sorenson defined the disease as a thoracic kyphosis, >45 degrees with 3 or more contiguous vertebrae wedged anteriorly a minimum of 5 degrees.[2] This definition is widely accepted as the diagnostic criteria for Scheuermann kyphosis.

Lateral radiographs show anterior wedging and narrowing of intervertebral disk spaces. There are often end-plate irregularities, termed Schmorl nodules (Figure 15-3). Schmorl nodules can be a common anomaly, occurring in 38% to 75% of the population and are not pathognomonic of the disease (Figure 15-4). The thoracic kyphotic Cobb angle should be measured and compared against a normal adolescent Cobb angle of 20 to 40 degrees. Spondylolysis and spondylolisthesis may be found with Scheuermann kyphosis and are associated with increased lumber lordosis.[13] Magnetic resonance imaging is not mandatory for evaluation except in the instance of patients presenting with neurologic deficits (Table 15-3).

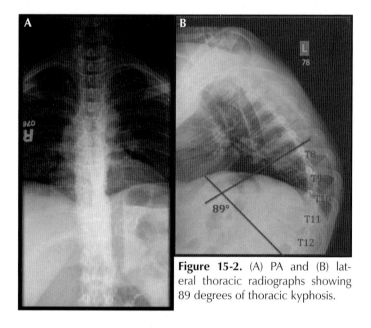

Figure 15-2. (A) PA and (B) lateral thoracic radiographs showing 89 degrees of thoracic kyphosis.

Figure 15-3. Magnified lateral radiograph showing anterior vertebral wedging and end-plate irregularities.

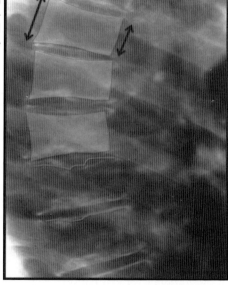

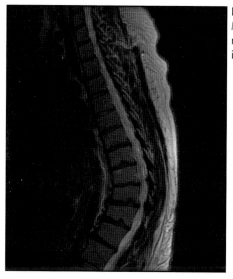

Figure 15-4. T1-weighted MRI showing Schmorl nodules and end-plate irregularities.

Table 15-3

IMAGING MODALITIES

Imaging Modality	Pertinent Image Views	Findings
Plain radiographs	Lateral view	≥3 consecutive wedged verte-brae with at least 5 degrees of wedging
Magnetic resonance imaging*	Coronal and sagittal views	Evaluate for disk herniation, tenting of spinal cord, extradural spinal cysts, and osteoporotic compression fractures

*When patient has neurologic findings on examination.

TREATMENT

The natural history of Scheuermann kyphosis is unclear. Considerations for treatment include the severity of deformity, age of the patient, and severity of the associated symptoms.[14]

Nonoperative Treatment

There is evidence to support observation, postural exercise, and sports for symptomatic treatment. Postural exercises increase flexibility and build paraspinal strength. Sports such as gymnastics, aerobics, swimming, basketball, and cycling that include hyperextension exercises should be encouraged.[15] Although there is no evidence of short- or long-term correction of deformities, there is evidence that patients experience pain reduction with exercise.

Bracing is reserved for skeletally patients with a kyphosis between 50 and 75 degrees, and a deformity apex between T6 and T9.[16] Patients who are compliant with bracing can achieve an initial 50% correction of their deformity, which can be largely maintained with full-time brace wear (>22 hours/day) for 12 to 18 months, followed by part-time wear (>12 hours/day) until skeletal maturity is reached. [6,17]

Adult patients receiving no surgical treatment have been found to have no difference from age-matched controls in ability to perform activities of daily living, days missed from work, social limitations, self-esteem, and pulmonary function tests. Patients without treatment are more likely to posses less physically demanding jobs.[13] In the adult patient complaining of pain, the root cause often is degenerative spondylosis. The mild scoliosis often associated with the kyphotic deformity in one-third of patients is usually nonprogressive in nature.

OPERATIVE TREATMENT

Patients with untreated severe disease may progress after skeletal maturity. Relative indications for surgical intervention include patients with kyphosis >70 degrees, adults with residual deformity and intractable pain, progressive neurologic deficits or progressive deformity, and skeletally immature patients who are poor bracing candidates.[6]

The 2 surgical treatment options include a posterior-only approach and a poster-anterior combined approach. The posterior-only approach is indicated in patients with a >70 degree kyphotic curve that corrects to at least 50 degrees preoperatively. The posterior approach offers decreased surgical time and avoids risks associated with a thoracotomy.[18] Patients after posterior-only surgery report subjective improvement in their pain symptoms that are unrelated to the ultimate change in their kyphosis.[19,20] A posterior-only approach can be complicated by hardware failure or hook pullout, pseudarthrosis, and loss of correction >5 degrees in long-term follow-up.[5] Patients with a rigid curve >70 degrees that does not correct preoperatively are good candidates for a combined approach. For these patients, an anterior release can make the curve more flexible.

An anterior release plus posterior instrumentation can be performed in 1 or 2 stages.[6] With the addition of an anterior approach, the anterior longitudinal ligament is released, complete diskectomies can be performed at multiple levels, bridging osteophytes can be excised, and bone grafting can be performed. Outcomes are similar to posterior releases, with patients reporting mild pain with rigorous activity but overall good improvement in pain that prior to surgery interfered with activities of daily living.[21]

CONCLUSION

The diagnosis of Scheuermann kyphosis is made on plain radiographs. The etiology of the disease process is unclear, and the natural history of the disease is controversial. Brace treatment appears to provide modest curve correction, but the efficacy is not conclusive. Surgery is associated with a high rate of complications, with most of the complications being major. However, surgery can improve kyphotic deformity, and there is evidence that it provides some pain relief.

REFERENCES

1. Scheuermann HW. The classic: kyphosis dorsalis juvenilis. *Clin Orthop Relat Res.* 1977;(128):5-7.

2. Sorenson KH. *Scheuermann's Juvenile Kyphosis: Clinical Appearances: Radiography, Aetiology and Prognosis.* Copenhagen, Denmark: Munkguard; 1964.
3. Wassmann K. Kyphosis juvenilis Scheuermann: an occupational disorder. *Acta Orthop Scand.* 1951;21(1-4):65-74.
4. Ascani E, Salsana V, Giglio G. The incidence and early detection of spinal deformities: a study based on the screening of 16,104 schoolchildren. *Ital J Orthop Traumatol.* 1977;3(1):111-117.
5. Wenger DR, Frick SL. Scheuermann kyphosis. *Spine (Phila Pa 1976).* 1999;24(24):2630-2639.
6. Raed M, et al. Scheuermann's kyphosis. *Curr Opin Pediatr.* 1999;11:70-75.
7. Yablon JS, Kasdon DL, Levine H. Thoracic cord compression in Scheuermann's disease. *Spine (Phila Pa 1976).* 1988;13(8):896-898.
8. Ali RM, Green DW, Patel TC. Scheuermann's kyphosis. *Curr Opin Pediatr.* 1999;11(1):70-75.
9. Damborg F, Engell V, Andersen M, Kyvik KO, Thomsen K. Prevalence, concordance and heritability of Scheuermann kyphosis based on study of twins. *J Bone Joint Surg Am.* 2006;88(10):2133-2136.
10. Findlay A, Conner AN, Connor JM. Dominant inheritance of Scheuermann's juvenile kyphosis. *J Med Genet.* 1989;26(6):400-403.
11. Ippilito E, Ponseti IV. Juvenile kyphosis: histological and histochemical studies. *J Bone Joint Surg Am.* 1981;63(2):175-182.
12. Bradford DS, Moe JH. Scheuermann's juvenile kyphosis: a histologic study. *Clin Orthop Relat Res.* 1975;(110):45-53.
13. Murray PM, Weinsten SL, Spratt KF. The natural history and long-term follow-up of Scheuermann kyphosis. *J Bone Joint Surg Am.* 1993;75(2):236-248.
14. Soo CL, Noble PC, Esses SI. Scheuermann kyphosis: long-term follow-up. *Spine J.* 2002;2(1):49-56.
15. Weiss HR, Dieckmann J, Gerner HJ. Effect of intensive rehabilitation on pain in patients with Scheuermann's disease. *Stud Health Technol Inform.* 2002;88:254-257.
16. Sachs B, Bradford D, Winter R, Lonstein J, Moe J, Willson S. Scheuermann's kyphosis: follow up of Milwaukee-brace treatment. *J Bone Joint Surg Am.* 1987;69(1):50-57.
17. Bradford DS, Moe JH, Montalvo FJ, Winter RB. Scheuermann's kyphosis and roundback deformity: results of Milwaukee brace treatment. *J Bone Joint Surg Am.* 1974;56(4):740-758.
18. Poolman RW, Been HD, Ubags LH. Clinical outcome and radiographic results after operative treatment of Scheuermann's disease. *Eur Spine J.* 2002;11(6):561-569.
19. Bradford, Moe JH, Montalvo FJ, Winter RB. Scheuermann's kyphosis: results of treatment by posterior spine arthrodesis in twenty-two patients. *J Bone Joint Surg Am.* 1975;57(4):439-448.
20. Papagelopoulos PJ, Klassen RA, Peterson HA, Dekutoski MB. Surgical treatment of Scheuermann's disease with segmental compression instrumentation. *Clin Orthop Relat Res.* 2001;(386):139-149.
21. Lowe TG, Kasten KD. An analysis of sagittal curves and balance after Cotrel-Dubousset instrumentation for kyphosis secondary to Scheuermann's disease: a review of 32 patients. *Spine (Phila Pa 1976).* 1994;19(15):1680-1685.

16

DISKITIS AND VERTEBRAL OSTEOMYELITIS OF THE SPINE

Greg Gebauer, MD and Alexander R. Vaccaro, MD, PhD

INTRODUCTION

Vertebral osteomyelitis, diskitis, and epidural abscesses form a spectrum of infectious processes that involve the spine. They can present as separate entities or in conjunction with one another. Diskitis is an infection that involves the intervertebral disk and is the most common spinal infection in pediatric patients. Vertebral osteomyelitis involves the vertebral body itself and is more common in older patients. Infection in the epidural space is termed an epidural abscess.

Rihn JA, Harris EB. *Musculoskeletal Examination of the Spine: Making the Complex Simple* (pp. 276-291).
© 2011 SLACK Incorporated.

This commonly occurs in association with either diskitis or vertebral osteomyelitis, although it can rarely occur as an isolated infection. Although relatively rare, recent studies suggest that the incidence of these infections may be on the rise.[1] An aging population, as well as the increasing prevalence of intravenous drug abuse, human immunodeficiency virus, and immunosuppressed patients may be contributing to this rise.

Infections of the spine can be challenging to diagnosis. Patients often have an indolent course of back or neck pain that, given the ubiquitous nature of these complaints, is either dismissed by the patient or physician. A high index of suspicion, a thorough examination of the patient, and an understanding of the factors that place a patient at risk for spinal infections are necessary to make the diagnosis. Mortality from spinal infections was historically as high as 25%,[2] although with modern antibiotics and treatment, this has decreased to 12%.[3] Patients with epidural abscesses or boney destruction and collapse related to the infection may present with neurologic compromise. This is especially true in patients with lesions of the cervical and thoracic spine, where the potential for neurologic injury is high. Swift recognition and treatment is necessary to prevent and treat such complications.

HISTORY

Pain is the most common presenting symptom of spinal infections, occurring in >90% of patients.[4] The pain can be either insidious in onset or occur acutely. Often, patients will have severe muscle spasms that contribute to their pain. Given the large differential diagnosis of back pain, the diagnosis of infection is often made late, up to 5 months after the onset of symptoms[5] (Table 16-1). The pain may or may not be worse at night or with activity. Patients may also complain of constitutional symptoms including fever, night sweats, generalized malaise, or weight loss. Depending on the location of the infection, patients may also present with complaints related to the neighboring structures; lumbar infections can cause irritation of the psoas muscles and cause pain with hip flexion, whereas cervical infections may cause retropharyngeal swelling, dysphagia, and respiratory difficulty. Patients with epidural extension or boney collapse may present with

Table 16-1

DIFFERENTIAL DIAGNOSIS OF SPINAL INFECTIONS

Neoplasm	Degenerative disc disease
Mechanical neck/back pain	Spondylothesis
Trauma	Abdominal pain
SI joint pain /infection	Charot arthropathy
Psoas abscess	

neurologic complaints, including radicular pain, weakness, numbness, or bowel and bladder dysfunction.

Infections can occur at any level but are most common in the lumbar spine, followed by the thoracic and cervical spine.[4] Several risk factors have been identified for development of spinal infections (Table 16-2). Typically, patients with spinal infections are >50 years, and infections more commonly occur in men.[6] Patients with diabetes, rheumatoid arthritis, immunosuppression, and a history of intravenous drug use as well as those who have recently undergone a spinal procedure or surgery are also at higher risk.

Children, especially younger ones, can be particularly hard to evaluate for spinal infections. Their complaints are often nonspecific, and younger children may be nonverbal. Patients and their parents may report increased irritability, abdominal pain, refusal to bear weight, or refusal to go into a flexed posture.

Patients who have had recent spinal procedures or surgery are also at risk for spinal infections. Most commonly, these patients present with increased back pain at the surgical site. Often, this pain does not improve over time and may be increasing in severity. Patients may also note drainage or irritation at the surgical site.

EXAMINATION

All patients suspected of having a spinal infection should undergo a thorough physical examination (Table 16-3). The

Table 16-2

RISK FACTORS FOR SPINAL INFECTION

Diabetes	Distant infection
HIV/AIDS	Malnutrition
Alcohol abuse	Renal disease
IV drug abuse	Smoking
Immunocompromise	Malignancy
Obesity	Age > 50 years old
Male gender	Trauma
Previous spinal surgery	Recent surgery or spinal procedure

Adapted from Eastlack RK, Kauffman CP. Pyogenic infections. In: Bono CM, Garfin SR, eds. *Orthopaedic Surgery Essentials: Spine*. Philadelphia, PA: Lippincott Williams & Wilkins; 2004.

neck and back should be assessed for tenderness and range of motion. A complete neurologic examination is then performed, with care taken to note any deficits. Long tract signs and reflexes should be tested. The presence or absence of a fever should be noted. Young children can be more difficult to examine but should be assessed for generalized irritability and refusal to bear weight.

After completing the physical examination, laboratory studies including a complete blood cell count, erythrocyte sedimentation rate (ESR), and C-reactive protein (CRP) should be obtained. ESR and CRP will be elevated in up to 90% of patients.[4] In addition, these markers can provide a benchmark from which the effectiveness of treatment can be assessed. A decreasing trend of these inflammatory markers is a reassuring sign that the treatment is effective. Although ESR can take weeks to rise in the face of infection or to normalize with adequate treatment, CRP responds more acutely and therefore is the preferred inflammatory marker for assessing response to treatment. Both of these factors can be elevated in the perioperative period, confounding the diagnosis in the patient who has had recent surgery; however, CRP should normalize between 5 and 10 days after surgery.[7] Elevated CRP after this

Table 16-3

METHODS FOR EXAMINING

- Physical exam findings
  - Tenderness over the area of the infection
  - Tenderness in the paraspinal muscles
  - Constitutional signs
    - Fever
    - Weight loss
    - Anorexia
- Laboratory data
  - Weight blood cell count
- Increased
  - Erythrocyete sedimentation rate
  - C-reactive protein
  - Blood cultures
    - May demonstrate causative organism
  - CT-guided biopsy
    - May demonstrate causative organism
- Radiographic findings
  - Plain X-rays
    - Boney destruction, sclerosis, disk space narrowing
  - CT scan
    - Boney destruction, sclerosis, disk space narrowing
    - May show abscesses/fluid collections
  - MRI
    - Decreased T1 and inreasde T2 signal
    - Peripheral enchancement with gadolinium
  - Bone scan
    - Increased uptake

Table 16-4

HELPFUL HINTS: DIAGNOSTIC ALGORITHM

Thorough history and physical	Care taken to note any risk factors and recent infections
Laboratory studies	CBC
	ESR
	CRP
	Blood cultures
Plain radiographs	
Advanced imaging	MRI
	CT scan
	Nuclear imaging
Biopsy	CT versus open; if needed blood cultures are negative

time period is concerning for infection. If patients have any underlying medical conditions or are at risk of being malnourished, laboratory studies appropriate to their medical condition should be ordered.

Blood cultures should be obtained on all patients who are believed to have a spinal infection. Blood cultures are positive in 24% to 59% of patients.[3,4] After an organism has been identified, it should be sent for sensitivity testing to help guide the selection of an appropriate antibiotic agent. Blood cultures are more likely to be positive if they are taken while patients are febrile. Table 16-4 provides a comprehensive algorithm for the evaluation of patients suspected of having a spinal infection.

PATHOANATOMY

Infections of the spine can occur from direct inoculation, by extension from adjacent infected structures, or through hematologic spread. Direct inoculation can occur from penetrating trauma or following a percutaneous or open surgical procedure. Even with proper antiseptic technique and prophylactic antibiotics, infections still occur in 4.2% of spine surgeries.[8] Higher incidences of infection have been associated with longer operative times and increased intraoperative blood loss, as well as in patients who have diabetes or are obese, and in those who have had previous surgical site infections. Patients who are undergoing surgery for trauma or who are malnourished may also be at increased risk for infection.

Most infections of the spine are thought to result from hematogenous spread, which can occur through either the arterial or venous systems. In adults, the infection is believed to begin in the vertebral end plate, eroding it, and spreading into the avascular nucleus pulposus. In children, the blood vessels cross the growth plate and enter the disk space, which may provide a route for direct infection of the disk space. After the infection has become established, it can spread to adjacent structures.

In the cervical spine, abscesses can spread anteriorly into the retropharyngeal space potentially causing swallowing and respiratory compromise. In the lumbar spine, the infection can spread into the retroperitoneal space and involve the psoas muscles. Posterior extension into the epidural space, especially in the cervical and thoracic spine, can compress the neural elements and potentially result in paralysis.

Staphylococcus aureus accounts for more than half of all bacterial spinal infections,[9] and more than half of these are methicillin-resistant strains.[10] Frequently, patients will have other sources of infection, including skin, urinary tract, or respiratory tract infections. Table 16-5 includes a list of pathogens and conditions with which they are commonly associated. Tuberculosis infections are relatively rare but do occur in immunocompromised patients and in patients who have spent time in certain third world countries. Spinal tuberculosis generally occurs via direct spread from the respiratory system into the thoracic spine, although it can occur in the lumbar or cervical spine. Tuberculosis osteomyelitis is different from typical bacterial infection in that there tends to be relative sparing of the disk space. Instead, the infection will commonly spread from one vertebral body to the next by forming granulomatous abscesses that travel under the anterior longitudinal ligament.

Table 16-5

COMMON PATHOGENS AND ASSOCIATED PATIENT CHARACTERISTCS

Organism	Patient Characteristics
Staph aureus	Most common organism
Pseudomonas	IV drug abuse
Salmonella	Sickle cell disease
Coagulase-negative Staph	Post-operative infection
Gram negative bacteria (commonly including E. coli, Klebsiella, and Proteus)	Genito-urinary infection
Tuberculous	Immunocompromise, history of travel to or residing in a 3rd world country

Adapted from Carragee E. Pyogenic vertebral osteomyelitis. *J Bone Joint Surg Am.* 1997;79(6):874-880; Sapico FL, Montgomerie JZ. Vertebral osteomyelitis. *Infect Dis Clin North Am.* 1990;4(3):539-550.

Chronic infections can lead to boney destruction, collapse of both the vertebral body and disk space, and progressive kyphotic deformity (Figure 16-1). Neurologic symptoms can occur either by impingement secondary to collapse of the vertebrae or due to posterior extrusion of fractured fragments into the epidural space. Extensive destruction may lead to instability of the spine.

IMAGING

Plain radiographs are the first imaging study that should be obtained in the evaluation of a patient with suspected spinal infection. Disk space collapse may be found in both diskitis and osteomyelitis, whereas boney destruction, sclerosis, and vertebral collapse are typically seen only in osteomyelitis (Table 16-6). It is important to remember, however, that findings on plain radiographs may lag behind the clinical course. Computed tomography (CT) may better define the extent of

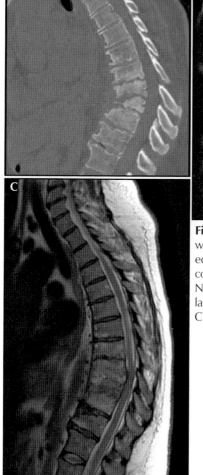
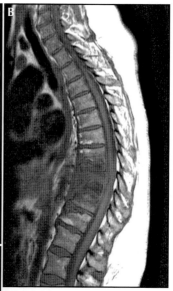

Figure 16-1. (A) Sagittal CT, (B) T1-weighted MRI, and (C) T2-weighted MRI showing late kyphotic collapse following osteomyelitis. Note the bony erosion and collapse are especially evident on CT.

boney destruction and aid in planning if surgery is indicated (Figure 16-2). CT may also help identify fluid collections around the spine. Computed tomography myelography is contraindicated in the setting of infection as the injection of the contrast dye into the thecal sac may introduce the infection into this area and lead to meningitis.

Table 16-6

IMAGING FEATURES

	Diskitis	Vertebral Osteomyelitis
Plain radiographs	Disk space narrowing	End plate and boney destruction, vertebral collapse, osteolysis and sclerosis
CT scans	Disk space narrowing	End plate and boney destruction, vertebral collapse, osteolysis and sclerosis
MRI	Low T1 signal, high T2 signal, peripheral enhancement with gadolinium	Low T1 signal, high T2 signal, peripheral enhancement with gadolinium
Nuclear medicine scans (Indium-111-labled WBC scans have a high false rate)	Increased uptake	Increased uptake

Adapted from Lederman HP, Schweitzer ME, Morrison WB, Carrino JA. MR imaging findings in spinal infections: rules or myths? *Radiology*. 2003;228(2):506-514.

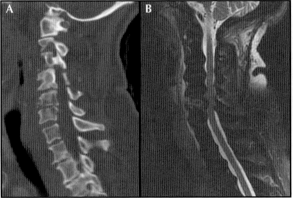

Figure 16-2. (A) Sagittal CT and (B) T2-weighted MRI of a C5-C6 osteodiskitis. Note the disk space collapse and boney erosions seen on CT. MRI reveals increased signal in the C5-C6 disk space and in the adjacent vertebral bodies.

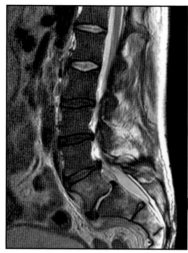

Figure 16-3. T2-weighted MRI showing L5-S1 diskitis. Note the increased signal in the L5-S1 disk space.

Magnetic resonance imaging (MRI) with contrast is the imaging study of choice for evaluation of the extent of spinal infection. Areas of infection will have increased signal on T2-weighted images and decreased signal on T1-weighted sequences (Figure 16-3).[10] The administration of gadolinium contrast will show peripheral enhancement. Contrast can also help differentiate infection from scar tissue in patients who have had previous surgery. MRI is helpful for identifying epidural abscesses and any compression of the neural elements (Figures 16-4 and 16-5).

The MRI appearance of infection in the spine can be similar to certain spinal tumors. Often, an infection may be distinguished from a tumor as tumors tend to spare the disk space, whereas infection commonly involves this area. The exception to this is tuberculosis, which typically spares the disk space but can cause disease in contiguous vertebral bodies. Care should be taken in distinguishing an infection from a tumor, and a biopsy may be necessary. After the diagnosis of infection has been made, an MRI of the entire spine should be considered to rule out any other foci of infection.

In patients unable to undergo MRI, nuclear medicine studies including technitium Tc 99m bone scans and indium In 111-labeled white blood cell scans may be considered. Bone scans are sensitive but not specific. Conversely, indium scans are highly specific but have a high false-negative rate. Using these studies in conjunction may help to improve accuracy.

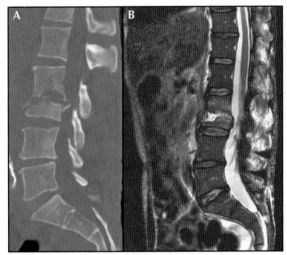

Figure 16-4. (A) Sagittal CT and (B) T2-weighted MRI show L2-L3 osteodiskitis. Note the collapse of the disk space and extrusion of a bony fragment into the spinal canal.

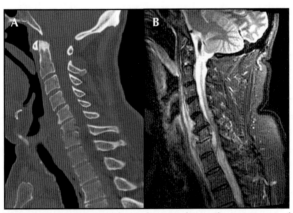

Figure 16-5. (A) Sagittal CT showing bony destruction and collapse at the C6-C7 level consistent with osteodiskitis. (B) Sagittal T2-weighted. *(continued)*

CT guided biopsy may be indicated for diagnostic purposes (ie, differentiating infection from tumor) as well as to obtain cultures for identifying causative organisms if blood cultures have been negative. If possible, antibiotics should be held prior

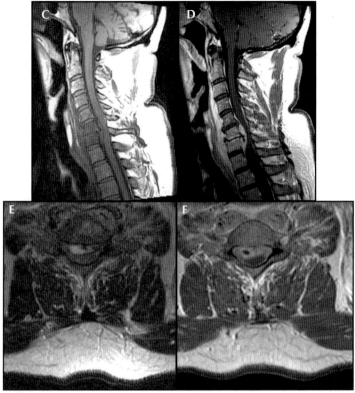

Figure 16-5 (continued). (C) T1-weighted, and (D) postcontrast T1-weighted MRI show a large epidural abscess extending posteriorly from the osteodiskitis and impinging on the spinal cord. Axial T2-weighted. (E) Postcontrast T1-weighted and (F) MRI images showing the same abscess.

to the biopsy to improve the likelihood of a positive culture result. CT-guided biopsies are successful in approximately 70% of cases.[12]

TREATMENT

The primary goals of treatment are to eliminate the infection, prevent and address any neurologic injury, maintain the structural integrity of the spine, and prevent deformity. Intravenous antibiotics should be administered for 2 to 6 weeks. The antibiotics should be tailored to the sensitivities of the identified organism. ESR and CRP levels should be followed serially to assess effectiveness of the treatment. After the course of intravenous antibiotics has been completed, patients are often transitioned to a course of oral antibiotics. The antibiotics selected, the route by which it is administered, and the duration of treatment depend on multiple factors, including the virulence of the organism, the health of the patient, and the presence of any retained hardware. An infectious disease consult should be considered to guide this treatment.

In addition to antibiotics, patients who are undergoing nonoperative treatment are generally placed in an appropriate brace or cervical collar. This may help to provide some comfort and pain relief as well as prevent the onset of late deformity. Any underlying medical conditions that may predispose the patient to infection should be addressed and optimized. The patient's nutritional status should be assessed and supplemented as needed to maximize the patient's ability to heal and clear the infection. A nutrition consult may be beneficial.

Surgery is indicated after a failure of nonoperative treatment to obtain a culture following failure of CT-guided biopsy, to address any neurologic compromise and to treat any deformity or instability related to the infection. Epidural abscesses are almost always treated with surgical decompression. These should be considered surgical emergencies in the setting of a neurologic deficit and treated with expeditious decompression and irrigation and débridement.[13]

Most infections of the spine are located anteriorly and therefore best addressed from an anterior approach. All infected material should be debrided and the epidural space decompressed if needed. Reconstruction of the spine is best performed using autogenous bone, either from the iliac crest or fibula. Vascularized bone grafts may be used, but this type of reconstruction is technically demanding. Allograft should be avoided if possible, given the concern with the infection continuing in the avascular bone. Structural titanium cages with nonstructural autograft have also been used with success.[14] A posterior approach may be appropriate in patients with epidural abscesses, particularly abscesses involving multiple levels (Figure 16-6). Anterior decompressions may require stabilization with posterior instrumentation to help correct deformity and provide stability.

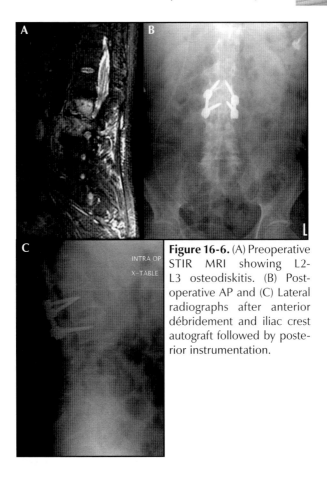

Figure 16-6. (A) Preoperative STIR MRI showing L2-L3 osteodiskitis. (B) Post-operative AP and (C) Lateral radiographs after anterior débridement and iliac crest autograft followed by posterior instrumentation.

CONCLUSION

Although relatively rare, infections of the spine can be difficult to diagnosis and to treat. A high clinical suspicion, careful physical examination, and appropriate laboratory and imaging studies are needed to make the diagnosis. Identification of the infectious pathogen is essential to guide antibiotic treatment. Surgery may be indicated to obtain a biopsy to help guide medical treatment, if medical management has failed or if the patient is developing instability, deformity, or neurologic compromise. Patients with epidural abscesses of the cervical and thoracic spine are at particular risk for neurologic injury and should be treated aggressively with surgical decompression. Serial CRP and ESR levels can be used to assess the effectiveness of treatment. Appropriate treatment of underlying medical conditions and maximization of the patient's nutrition will further aid in healing and clearing of the infection.

REFERENCES

1. Acosta FL Jr, Galvez LF, Aryan HE, Ames C. Recent advances: infections of the spine. *Curr Infect Dis Rep.* 2006;8(5):390-393.
2. Guri JP. Pyogenic osteomyelitis of the spine: differential diagnosis through clinical and roentgenographic observations. *J Bone Joint Surg Am.* 1946;28(1):29-39.
3. Carragee E. Pyogenic vertebral osteomyelitis. *J Bone Joint Surg Am.* 1997;79(6):874-880.
4. Sapico FL, Montgomerie JZ. Pyogenic vertebral osteomyelitis: report of nine cases and review of the literature. *Rev Infect Dis.* 1979;1(5):754-776.
5. Perronne C, Saba J, Behloul Z, et al. Pyogenic and tuberculous spondylodiskitis (vertebral osteomyelitis) in 80 adult patients. *Clin Infect Dis.* 1994;19(4):746-750.
6. Eastlack RK, Kauffman CP. Pyogenic infections. In: Bono CM, Garfin SR, eds. *Orthopaedic Surgery Essentials: Spine*, Philadelphia, PA: Lippincott Williams & Wilkins; 2004:73-80.
7. Thelander U, Larsson SL. Quantitation of C-reactive protein levels and erythrocyte sedimentation rate after spinal surgery. *Spine (Phila Pa 1976).* 1992;17(4):400-404.
8. Pull ter Gunne AF, Cohen DB. Incidence, prevalence, and analysis of risk factors for surgical site infection following adult spinal surgery. *Spine (Phila Pa 1976).* 2009;34(13):1422-1428.
9. Hadjipavlou AG, Mader JT, Necessary JT, Muffoletto AJ. Hematogenous pyogenic spinal infections and their surgical management. *Spine (Phila Pa 1976).* 2000;25(13):1668-1678.
10. Livorsi DJ, Daver NG, Atmar RL, Shelburne SA, White AC Jr, Musher DM. Outcomes of treatment for hematogenous *Staphylococcus aureus* vertebral osteomyelitis in the MRSA era. *J Infect.* 2008;57(2):128-131.
11. Lederman HP, Schweitzer ME, Morrison WB, Carrino JA. MR imaging findings in spinal infections: rules or myths? *Radiology.* 2003;228(2):506-514.
12. Kornblum MB, Wesolowski DP, Fischgrund JS, Herkowitz HN. Computed tomography-guided biopsy of the spine: a review of 103 patients. *Spine (Phila Pa 1976).* 1998;23(1):81-85.
13. Sa mpath P, Rigamonti D. Spinal epidural abscess: a review of epidemiology, diagnosis, and treatment. *J Spinal Disord.* 1999;12(2):89-93.
14. Korovessis P, Petsinis G, Koureas G, Iliopoulos P, Zacharatos S. One-stage combined surgery with mesh cages for treatment of septic spondylitis. *Clin Orthop Relat Res.* 2006;444:51-59.

17

PRIMARY BONE TUMORS OF THE SPINE

Nelson S. Saldua, MD and James S. Harrop, MD, FACS

INTRODUCTION

More than 90% of tumors that involve the spinal column are of metastatic origin.[1] Primary spinal osseous neoplasms are rare entities that are infrequently encountered by spine surgeons, yet they must be considered in the differential diagnosis when treating a patient with a spinal column lesion.

Regardless of the origin of the spinal column lesion, the goals of treatment remain the same. Management includes obtaining a definitive tissue diagnosis from either biopsy or excision, and tailoring treatment according to the tumor type, histologic grade, and the patient's individual presentation. In

Rihn JA, Harris EB. *Musculoskeletal Examination of the Spine: Making the Complex Simple* (pp. 292-312).
© 2011 SLACK Incorporated.

addition, treatment is further focused on preservation of neurologic function and maintenance of spinal column stability.

HISTORY

The most common chief complaint and presenting symptom of a patient with a primary spinal column lesion is pain. Pain is reported in approximately 85% of patients, with the majority of the pain being axial, although some patients report radicular symptoms due to neural compression.[2] Midline axial pain, whether located in the cervical, thoracic, or lumbar region, is rather nonspecific and can have a wide differential diagnosis. Pain from a neoplastic process can be differentiated from other causes by the quality of the pain. Pain from a neoplastic process in the spinal column is typically progressive and unrelenting. Night pain should be a red flag to the clinician, since it is classically described as being due to an oncologic process. Unlike mechanical pain seen with degenerative disease, this pain is not usually exacerbated with activity and ameliorated with rest. The clinician must be aware of other systemic symptoms of neoplasm, such as fevers, chills, night sweats, and unintended weight loss. These systemic symptoms are often encountered with osseous tumors such as lymphoma, myeloma, Ewing's sarcoma, and metastatic lesions.

Severe pain associated with a bone neoplasm is usually due to bone remodeling, and acute pain may be due to a pathologic fracture. In addition, pain may be due to stretching of the overlying periosteum. Radicular symptoms or neuropathic pain occur when local tumor extension causes compression or invasion of nearby nerve roots.

In some cases of benign tumors affecting the posterior elements, such as osteoid osteomas and osteoblastomas, scoliosis can occur.[3] In contrast to the painless deformity typically encountered with adolescent idiopathic scoliosis, the scoliosis or deformity resulting from these posterior element lesions is often associated with a significant amount of pain from paraspinal muscle spasms and stiffness.

EXAMINATION

The findings on the physical examination of a patient with a spinal column lesion are often variable (Table 17-1). On one end of the spectrum, a slowly progressive benign lesion or any lesion detected early in its presentation may show no focal abnormality on the physical as well as the neurologic examination. However, on the other end of the spectrum, patients with spinal cord or nerve root compression may show a focal neurologic deficit. Depending on the site of compression, patients may exhibit signs and symptoms of myelopathy (including hyperactive deep tendon reflexes; Hoffmann's, Romberg's, or Babinski's signs; clonus; and inability to tandem gait), radiculopathy, bowel or bladder dysfunction, paraparesis, or even tetraparesis.[4] With this wide variety of physical examination findings, it is crucial for the clinician to consider the patient's overall presentation. Any focal neurologic finding must be correlated with the proper imaging studies to make the correct diagnosis.

Primary bone tumors of the spine are rarely felt as a palpable mass on examination due to the overlying muscle and skin. The most likely exception to this statement is that of a sacral chordoma. Sacral chordomas grow anteriorly compressing the rectal vault and can often present with bowel and bladder dysfunction as well as a palpable mass on rectal examination.[5]

PATHOANATOMY

Primary tumors of the spinal column can be either benign or malignant. Malignancies, typically metastatic in nature, most often arise in the anterior columns of older patients, whereas benign lesions occur in the posterior elements and are typically seen in the pediatric population. The most common primary osseous benign spinal tumors include osteochondroma, osteoid osteoma, osteoblastoma, aneurysmal bone cysts, hemangioma, giant cell tumor, and eosinophilic granuloma (Table 17-2). The more aggressive primary malignancies that are found in the spinal column include osteogenic sarcoma, chondrosarcoma, Ewing's sarcoma, chordoma, multiple myeloma, and lymphoma.

Table 17-1

METHODS FOR EXAMINATION

- Diagnostic imaging includes plain radiographs, CT scans, MRI scans and bone scans. Plain radiographs may be normal unless significant bony destruction has occurred or unless a structural abnormality has been caused. CT scans help delineate bony anatomy and quantify bony destruction, while MRI can help show soft tissue extension of any mass. Bone scans have the advantage of detecting lesions throughout the body.
- Physical examination must always correlate with the radiographic work-up. A complete neurological exam includes an assessment of motor strength, a dermatome-based sensory examination, and an assessment of deep tendon reflexes. Digital rectal examination is needed for assessment of sacral nerve function.
- Common muscle strength tests that correlate with particular nerve roots:
  - C5: Shoulder abduction (deltoid)
  - C6: Elbow flexion and wrist extension (biceps and wrist extensors)
  - C7: Elbow extension and wrist flexion (triceps and wrist flexors)
  - C8: Hand grip (finger flexors)
  - T1: Finger extension and finger abduction/adduction (hand intrinsics)
  - L1: Hip flexion (iliopsoas)
  - L2-L3: Knee flexion (hamstrings)
  - L3-4: Knee extension (quadriceps)
  - L4: Ankle dorsiflexion (tibialis anterior)
  - L5: Great toe extension (EHL)
  - S1: Ankle platarflexion (gastrocnemius and soleus)
- Sensory dermatomes to remember include:
  - C5: Lateral arm
  - C6: Lateral forearm and hand
  - C7: Middle finger
  - C8: Medial forearm and hand
  - T1: Medial arm
  - T4: Nipple
  - T10: Umbilicus
  - L1: Inguinal crease
  - L4: Anterior thigh and medial leg
  - L5: Lateral thigh and dorsum of foot
  - S1: Posterior thigh and leg, plantar surface of foot

Table 17-2

COMMON PRIMARY OSSEOUS BENIGN AND MALIGNANT SPINAL TUMORS

Type of Tumor	Pathognomonic Feature or Characteristic Finding
Osteochondroma	Bony projection can occasionally be palpated; cortical bone lesion is in continuity with the adjacent normal cortical bone; medullary canal of lesion is also in continuity with medullary canal of underlying bone (Figure 17-1)
Osteoid osteoma	Night pain typically relieved by nonsteroidal anti-inflammatory drugs; central nidus seen on imaging, particularly axial CT (Figure 17-2)[6]
Osteoblastoma	Also with central nidus but entire lesion is >2 cm (Figure 17-3)[7]
Aneurysmal bone cyst	Lytic lesion that expands the posterior elements[8]; air-fluid levels can be seen on CT or MRI (Figure 17-4)[9]
Hemangioma	Jailhouse vertebral body resulting from destruction of horizontal trabeculae and preservation of vertically oriented trabeculae (Figures 17-5 and 17-6)
Giant cell tumor	Expansile lytic lesion adjacent to the subchondral bone; commonly affects the sacrum[10,11]
Eosinophilic granuloma	Vertebra plana (Figure 17-7)
Osteosarcoma	Poorly defined, aggressive lesion
Chondrosarcoma	Cartilage-forming lesion with characteristic appearance of rings and whirls on plain radiographs (Figure 17-8)
Ewing's sarcoma	Small round blue cells on histology
Chordoma	In cases of sacral chordoma, the mass often produces continuous rectal pain; mass can sometimes be palpated on rectal examination (Figure 17-9)
Multiple myeloma	Small round blue cells on histology
Lymphoma	May be epidural, in bone marrow, or systemic

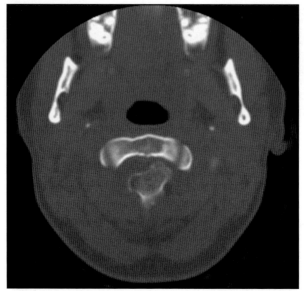

Figure 17-1. Osteochondroma within the spinal canal showing cortical bone and medullary continuity with the vertebrae.

IMAGING

Plain radiographs of the region of the spinal column in question are typically the first imaging studies obtained. There should be at least 2 views of the spine in orthogonal planes as lesions will often only be apparent on one radiographic view. Because there must be a great deal of destruction of the vertebral body's trabecular bone (approximately 50%) before this is radiographically apparent, plain radiographs are often normal.[12]

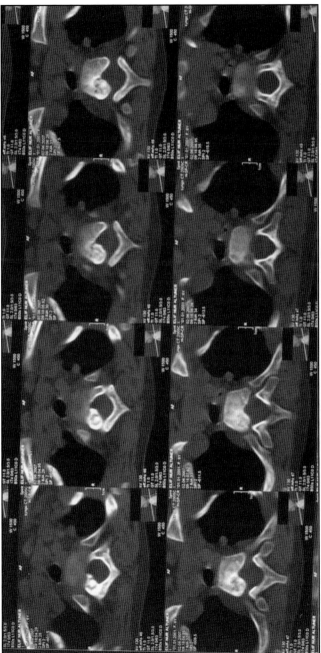

Figure 17-2. Axial CT scan images showing central nidus often seen in osteoid osteoma.

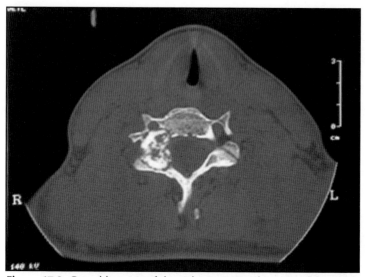

Figure 17-3. Osteoblastoma of the right posterior elements of the cervical spine.

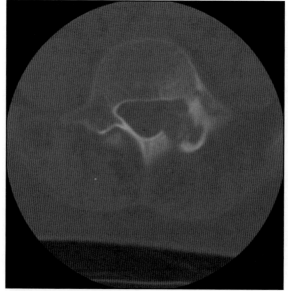

Figure 17-4. Aneurysmal bone cyst of the posterior elements of the lumbar spine.

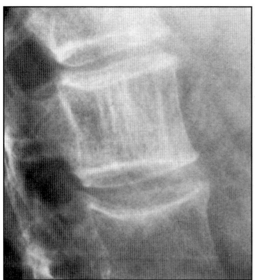

Figure 17-5. Lateral radiograph of the spine showing the "jailhouse vertebra" typically seen with hemangioma of the vertebral body.

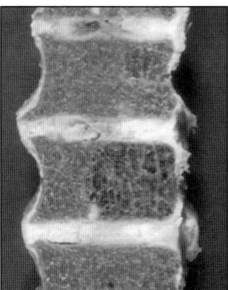

Figure 17-6. Gross specimen of a vertebral hemangioma. Note the preferential destruction of the horizontally-oriented trabeculae and preservation of the vertically-oriented trabeculae which gives the jailhouse vertebral appearance.

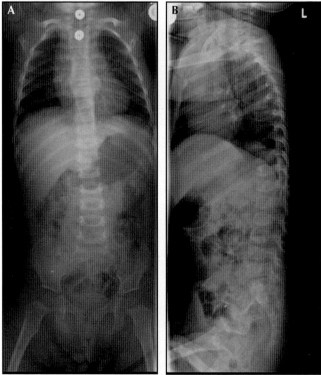

Figure 17-7. Anterposterior and lateral radiographs showing the flattened vertebra, or vertebral plana, seen in eosinophilic granuloma.

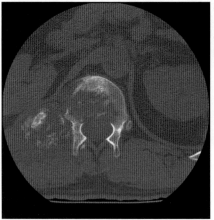

Figure 17-8. Axial CT scan showing characteristic appearance of a cartilaginous tumor of the spine.

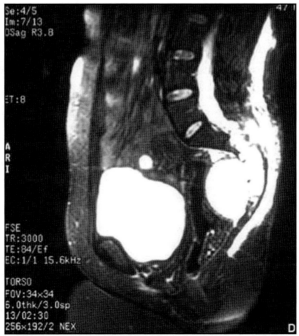

Figure 17-9. Sagittal MRI showing a sacral mass. Differential diagnosis includes chordoma versus giant cell tumor.

With metastatic disease, there is typically destruction of the anterolateral portion of the spinal column. This lesion typically affects one pedicle and on an anteroposterior radiographic yields the classic "winking owl" sign (Figure 17-10). The vertebral body represents the head of the owl, and the intact pedicle is the open eye of the owl. The spinous process comprises the beak, and the destroyed pedicle is the "winking eye."[13] The radiographic differential for these lesions also includes infection. These lesions can usually be differentiated from infection because neoplasms tend to preserve the intervertebral disk, whereas infections will involve the disk, resulting in disk height loss and sclerosis. The exception to this is tuberculosis, which often spares the disk space despite being an infection.

Bone scans have the advantage of evaluating the entire body and can serve as a screening tool secondary to an ability to detect lesions as small as 2 cm (Figure 17-11). Computed

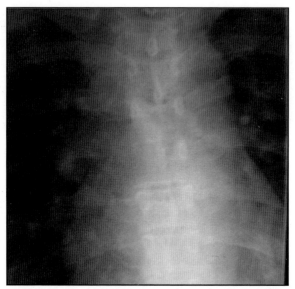

Figure 17-10. Anterposterior radiograph showing the "winking owl" sign characteristic from unilateral pedicle destruction.

tomography (CT) scans are the optimal study to determine osseous involvement and destruction with a neoplastic process as well as structural integrity of the spine. Myelography is an invasive procedure and has been used to determine cord or nerve root compression but has been largely replaced with magnetic resonance imaging (MRI). MRI is the optimal imaging modality for the neural elements since it is noninvasive and provides detailed imaging of the spinal cord parenchyma. Myelography remains a useful imaging alternative when there is a contraindication to obtaining an MRI, such as the presence of a cardiac pacemaker, body habitus, or the patient's inability to remain still in the MRI scanner.

MRI can be used to delineate the full extent of the lesion including the amount of soft tissue extension. It may also show the presence of any additional lesions, whereas contrast-enhanced MRI can be used to monitor responses to chemotherapy as well as possible recurrence. Caution must be taken, in the case of benign osseous tumors, as there is evidence that MRI may be hypersensitive in determining the degree of aggressiveness of the lesion.

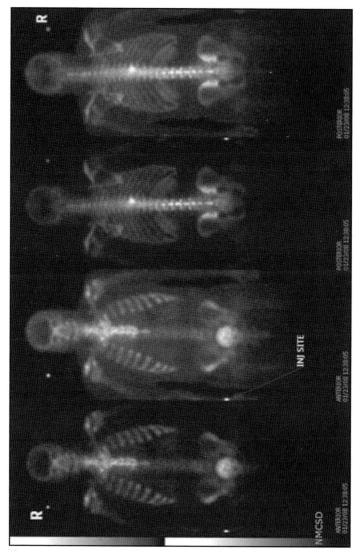

Figure 17-11. Bone scan showing increased signal corresponding with a mid-thoracic spine lesion.

TREATMENT

There are currently 2 staging systems used to classify spine tumors and guide treatment. The Enneking classification was

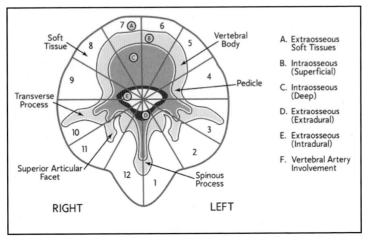

Figure 17-12. The 12 segments of the spine as described by the WBB classification. (Reprinted with permission of Chan P, Boriani S, Fourney DR, et al. An assessment of the reliability of the Enneking and Weinstein-Boriani-Biagini classifications for staging of primary spinal tumors by the Spine Oncology Study Group. *Spine (Phila Pa 1976)*. 2009;34(4): 384-391.)

initially introduced to guide the treatment of appendicular musculoskeletal tumors. This classification is based on local extent of tumor progression, histologic grade, and presence of metastasis.[14-16] Extrapolation of this classification system to the axial skeleton does not take into account the presence of a continuous epidural compartment, the neurologic deficits from resection of neural elements, and the need for preserving or restoring spinal stability.

Therefore, the Weinstein-Boriani-Biagini (WBB) classification system was proposed to deal with these added concerns with treatment of tumors of the vertebral column.[17] The WBB system divides the vertebral body in the axial plane into 12 radiation zones, which are numbered 1 to 12 similar to a clock face. It then also divides the vertebral body into 5 concentric layers based on bony and dural involvement. The vertebrae involved also define the extent of the tumor. The WBB staging system has recently been modified according to a consensus paper from the Spine Oncology Study Group (Figure 17-12).[18] Whether the Enneking or WBB system is used, the goal is to achieve the appropriate surgical margin while at the same time avoiding high-morbidity surgery.

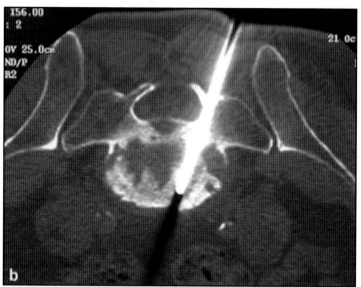

Figure 17-13. CT guided biopsy of the sacrum.

Primary spinal column lesions can be malignant or benign, with malignant lesions most often affecting the vertebral body and possibly one or both pedicles secondary to extension from the anterior component. Benign lesions, however, are more likely to arise in the posterior elements.[2] To make a histologic diagnosis, which will in turn help guide treatment, a tissue biopsy should be performed. Options include image-guided (using fluoroscopy or CT) percutaneous needle biopsy as well as open biopsy (Figure 17-13). Basic tumor principles should be adhered to in the event that the biopsy reveals the lesion to be malignant. The definitive resection and reconstruction must always be kept in mind when planning and performing the initial biopsy.

After a tissue diagnosis is established, treatment is tailored to the specific type of tumor, the histologic grade, and its impact on spinal stability. Treatment of benign spinal tumors is usually through surgical excision (Table 17-3).[1] Regardless of histologic grade or tumor type, the overall stability of the spine must be addressed when considering surgical management. In cases in which spinal stability is not compromised, the patient may be managed nonoperatively. In these

Table 17-3

TREATMENT OF PRIMARY BONE TUMORS OF THE SPINE

Type of Tumor	Treatment
Osteochondroma	Wide surgical excision if symptomatic; otherwise nonoperative management can be considered
Osteoid osteoma	Nonsteroidal anti-inflammatory drugs, radiofrequency ablation, or surgical resection[19]
Osteoblastoma	Surgical resection
Aneurysmal bone cyst	CT-guided injection of sclerotic solution versus preoperative embolization andresection[20-22]
Hemangioma (aggressive)	Frequently responds to radiation; preoperative embolization and surgical resection or reconstruction are considered for pathologic fractures
Giant cell tumor	Wide excision with adjuvants and reconstruction[23,24]
Eosinophilic granuloma	Self-limiting natural history; management observational versus corticosteroid injection
Osteosarcoma	Wide resection, decompression, and fusion
Chondrosarcoma	Wide resection, decompression, and fusion
Ewing's sarcoma	Wide resection, decompression, and fusion
Chordoma	Wide resection, decompression, and fusion
Multiple myeloma	Radiation therapy and systemic control
Lymphoma	Radiation therapy and systemic control

cases, treatment options vary from symptomatic management to bracing. When spinal stability is affected, surgical management is considered if the patient is a surgical candidate. For those lesions that cannot be treated nonoperatively, the treatment typically includes wide resection with negative margins if anatomically possible. If it is not possible to achieve negative margins, then a debulking procedure can be performed in which the surgeon removes as much of the tumor as possible.

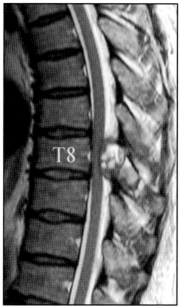

Figure 17-14. Sagittal MRI showing spinal cord compression from spinal lesion.

Decompression of the spinal canal should be performed if there is compression on the spinal cord, particularly if there is a progressive neurologic deficit (Figure 17-14). Reconstruction after the resection often includes the placement of anterior cages or bone graft and posterior instrumentation (Figure 17-15).[2,25,27]

A preoperative embolization by an interventional radiologist can be considered when surgery on a well-vascularized tumor is considered (Figure 17-16). This is commonly used in the surgical management of hemangiomas, aneurysmal bone cysts, and renal cell carcinoma metastases.[20,28-30] Preoperative embolization has also been described as an adjunct to surgical management of osteoblastomas.[7,31] In certain cases, minimally invasive procedures such as radiofrequency ablation for osteoid osteomas or CT-guided injections for aneurysmal bone cysts can be performed by an interventional radiologist.

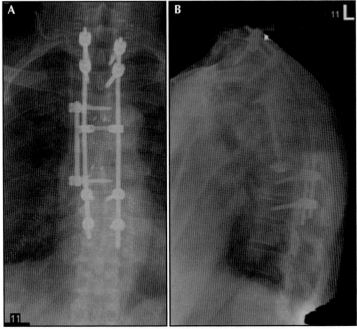

Figure 17-15. Postoperative radiographs showing reconstruction and stabilization after resection of the spinal lesion seen in Figure 17-14.

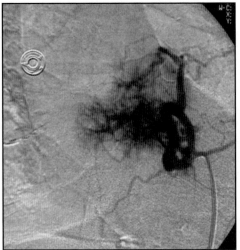

Figure 17-16. Preoperative embolization of highly-vascularized tumors is often needed to help minimize intraoperative blood loss.

Table 17-4

HELPFUL HINTS

- Primary bone tumors occur much less commonly than metastatic lesions in the spine. Metastatic lesions must be considered in the differential diagnosis of any lesion of the spine.
- Tumors within the spine, either primary or metastatic, can be confused with infections on diagnostic imaging. Infectious etiologies must be properly worked up with cultures of biopsied material.
- Certain lesions have a predilection for certain portions of the spine. Metastatic lesions and hemangiomas tend to occur within the vertebral body. Lesions that typically occur within the posterior elements include osteoid osteomas and osteoblastomas.

The treatment for malignant lesions requires a multidisciplinary team approach. Radiation, chemotherapy, or both can be used preoperatively as well as postoperatively. In the case of osteosarcoma, neoadjuvant and adjuvant chemotherapy increases survival rates and decreases recurrence rates. Radiation therapy, while having the advantageous result of causing tumor cell death, can also lead to deleterious effects on normal tissue. This must be taken into consideration when electing to treat a patient with perioperative radiation as this can have consequences with regard to surgical wound healing.

CONCLUSION

Primary bone tumors are uncommonly encountered in the spine (Table 17-4). Due to their rarity, these lesions are frequently overlooked in the differential diagnosis. Unfortunately, the variable presentations, physical examination findings, and radiographic imaging features make the diagnosis of a primary bone tumor of the spine difficult. All of these variables must be correlated with one another to make the correct diagnosis and manage the patient appropriately. It is important for clinicians to include primary bone tumors of spine within their differential diagnosis to correctly manage this rare subset of patients.

REFERENCES

1. Simmons ED, Zheng Y. Vertebral tumors: surgical versus nonsurgical treatment. *Clin Orthop Relat Res.* 2006;443:233-247.
2. Weinstein JN, McLain RF. Primary tumors of the spine. *Spine (Phila Pa 1976).* 1987;12(9):843-851.
3. Saifuddin A, White J, Sherazi Z, Shaikh MI, Natali C, Ransford AO. Osteoid osteoma and osteoblastoma of the spine. Factors associated with the presence of scoliosis. *Spine (Phila Pa 1976).* 1998;23(1):47-53.
4. Sciubba DM, Petteys RJ, Garces-Ambrossi GL, et al. Diagnosis and management of sacral tumors. *J Neurosurg Spine.* 2009;10(3):244-256.
5. Samson IR, Springfield DS, Suit HD, Mankin HJ. Operative treatment of sacrococcygeal chordoma: a review of twenty-one cases. *J Bone Joint Surg Am.* 1993;75(10):1476-1484.
6. Kan P, Schmidt MH. Osteoid osteoma and osteoblastoma of the spine. *Neurosurg Clin N Am.* 2008;19(1):65-70.
7. Denaro V, Denaro L, Papalia R, Marinozzi A, Di Marino A. Surgical management of cervical spine osteoblastomas. *Clin Orthop Relat Res.* 2007;455:190-195.
8. Burch S, Hu S, Berven S. Aneurysmal bone cysts of the spine. *Neurosurg Clin N Am.* 2008;19(1):41-47.
9. Caro PA, Mandell GA, Stanton RP. Aneurysmal bone cyst of the spine in children: MRI imaging at 0.5 tesla. *Pediatr Radiol.* 1991;21(2):114-116.
10. Hosalkar HS, Jones KJ, King JJ, Lackman RD. Serial arterial embolization for large sacral giant-cell tumors: mid- to long-term results. *Spine (Phila Pa 1976).* 2007;32(10):1107-1115.
11. Leggon RE, Zlotecki R, Reith J, Scarborough MT. Giant cell tumor of the pelvis and sacrum: 17 cases and analysis of the literature. *Clin Orthop Relat Res.* 2004;(423):196-207.
12. Edelstyn GA, Gillespie PJ, Grebbell FS. The radiological demonstration of osseous metastases: experimental observations. *Clin Radiol.* 1967;18(2):158-162.
13. Jacobson HG, Poppel MH, Shapiro JH, Grossberger S. The vertebral pedicle sign: a roentgen finding to differentiate metastatic carcinoma from multiple myeloma. *Am J Roentgenol Radium Ther Nucl Med.* 1958;80(5):817-821.
14. Enneking WF, Spanier SS, Goodman MA. Current concepts review: the surgical staging of musculoskeletal sarcoma. *J Bone Joint Surg Am.* 1980;62(6):1027-1030.
15. Enneking WF, Spanier SS, Goodman MA. A system for the surgical staging of musculoskeletal sarcoma. *Clin Orthop Relat Res.* 1980;(153): 106-120.
16. Enneking WF. A system of staging musculoskeletal neoplasms. *Clin Orthop Relat Res.* 1986;(204):9-24.
17. Boriani S, Weinstein JN, Biagini R. Primary bone tumors of the spine: terminology and surgical staging. *Spine (Phila Pa 1976).* 1997; 22(9): 1036-1044.

18. Chan P, Boriani S, Fourney DR, et al. An assessment of the reliability of the Enneking and Weinstein-Boriani-Biagini classifications for staging of primary spinal tumors by the Spine Oncology Study Group. *Spine (Phila Pa 1976)*. 2009;34(4):384-391.

19. Ozaki T, Liljenqvist U, Hillmann A, et al. Osteoid osteoma and osteoblastoma of the spine: experiences with 22 patients. *Clin Orthop Relat Res.* 2002;397:394-402.

20. Guibaud L, Herbreteau D, Dubois J, et al. Aneurysmal bone cysts: percutaneous embolization with an alcoholic solution of zein–series of 18 cases. *Radiology.* 1998;208(2):369-373.

21. Liu JK, Brockmeyer DL, Dailey AT, Schmidt MH. Surgical management of aneurysmal bone cysts of the spine. *Neurosurg Focus.* 2003;15(5):E4.

22. Boriani S, De Iure F, Campanacci L, et al. Aneurysmal bone cyst of the mobile spine: report on 41 cases. *Spine (Phila Pa 1976).* 2001;26(1):27-35.

23. Ozaki T, Liljenqvist U, Halm H, Hillmann A, Gosheger G, Winkelmann W. Giant cell tumor of the spine. *Clin Orthop Relat Res.* 2002;(401): 194-201.

24. Fidler MW. Surgical treatment of giant cell tumours of the thoracic and lumbar spine: report of nine patients. *Eur Spine J.* 2001;10(1):69-77.

25. Tomita K, Kawahara N, Baba H, Tsuchiya H, Fujita T, Toribatake Y. Total en bloc spondylectomy: a new surgical technique for primary malignant vertebral tumors. *Spine (Phila Pa 1976).* 1997;22(3):324-333.

26. McAfee PC, ZdeblickTA. Tumors of the thoracic and lumbar spine: surgical treatment via the anterior approach. *J Spinal Disord.* 1989;2(3): 145-154.

27. Sundaresan N, Steinberger AA, Moore F, et al. Indications and results of combined anterior-posterior approaches for spine tumor surgery. *J Neurosurg.* 1996;85(3):438-446.

28. Green JA, Bellemore MC, Marsden FW. Embolization in the treatment of aneurysmal bone cysts. *J Pediatr Orthop.* 1997;17(4):440-443.

29. De Cristofaro R, Biagini R, Boriani S, et al. Selective arterial embolization in the treatment of aneurysmal bone cyst and angioma of bone. *Skeletal Radiol.* 1992;21(8):523-527.

30. DeRosa GP, Graziano GP, Scott J. Arterial embolization of aneurysmal bone cyst of the lumbar spine: a report of two cases. *J Bone Joint Surg Am.* 1990;72(5):777-780.

31. Trubenbach J, Nagele T, Bauer T, Ernemann U. Preoperative embolization of cervical spine osteoblastomas: report of three cases. *AJNR Am J Neuroradiol.* 2006;27(9):1910-1912.

18

METASTATIC DISEASE OF THE SPINE

Jeffrey M. Tuman, MD; Matthew R. Eager, MD;
and Adam L. Shimer, MD

INTRODUCTION

Approximately 1.4 million new cases of cancer are diagnosed in the United States each year, and 500,000 deaths are attributable to the disease. Primary malignances such as breast, lung, thyroid, renal, and prostate have the propensity to metastasize to bone.[1] Bone metastases are the most frequent source of severe pain in patients with advanced cancer.[2] The spine is the most common osseous site of metastatic disease, with 20,000 new cases per year.[3] Metastatic lesions of the spine are 10 times more common than primary bone tumors. Approximately 20% to 30% of patients will have symptomatic spinal metastasis in their disease

Rihn JA, Harris EB. *Musculoskeletal Examination of the Spine: Making the Complex Simple* (pp. 313-324).
© 2011 SLACK Incorporated.

course; however, at death, >90% have lesions. Careful attention to atypical symptoms presenting with back pain is crucial since approximately 12% to 20% of new diagnoses are made with patients presenting with symptomatic spinal metastases.

Treatment goals include pain relief and maintenance of spinal and neurologic stability. Most metastatic lesions can be successfully treated medically with a combination of external beam radiation therapy (for radiation-sensitive primary tumors), chemotherapy, and a pain medication regimen. Surgical intervention may be required if medical treatment is not effective for pain control, if the spine is mechanically unstable, and if surgical stabilization or decompression is required to prevent or treat neurologic deterioration.

HISTORY

A thorough history remains critical to a correct and expeditious diagnostic process. The most common presenting complaint of spinal metastases is localized pain (95%).[4] In fact, axial pain is the most common complaint in patients with unknown metastatic disease of the spine. Associated pain is typically nonmechanical and unrelieved by medication or other standard treatment modalities. Less commonly, expansile lesions can also encroach on neural elements, yielding neurologic symptoms (5%).[5] Nerve root compression will usually result in dermatomal radicular-type symptoms (pain, numbness, paresthesias, and weakness) and is dependent on the level of direct compression. Spinal cord or thecal sac compression, usually from gross deformity or direct compression from the tumor, may result in myelopathic complaints including gait ataxia, poor manual dexterity, and spasticity. Progressive urinary or bowel dysfunction and perineal numbness are alarming symptoms of possible cauda equina pathology. Pain predates neurologic symptoms on average by 7 months.[6]

A current or previous history of extramusculoskeletal malignancy should raise obvious concern. Any spine examination should include a thorough medical and family history, including cancer. Breast cancer in particular can represent 15 to 20 years after remission, detected as a bony metastasis. If a suspicious lesion is identified, a focused history can be completed in an effort to identify the primary source. This includes symptoms such as hemoptysis and shortness of breath (lung carcinoma), hematuria (renal cell carcinoma), or urinary dysfunction (prostate). Any patient complaining of spine pain or neurologic symptoms should also be queried regarding systemic "red flag" symptoms. Recent unintended weight loss, loss of appetite, fatigue, fevers, chills, night sweats, and easy bruising are all classic constitutional complaints that may accompany a metastatic condition. Cancer pain should be suspected if there is a prominent nocturnal characteristic to the patient's complaints.

A pathologic fracture secondary to metastatic disease must always be considered anytime a low-energy fracture of the spine exists. Most commonly, this presents as a compression-type fracture. This is especially true in a patient with a cancer history and otherwise low risk of osteoporosis, such as a man with lung cancer. Any pain preceding a compression fracture should also be thoroughly investigated for pathology.

EXAMINATION

A neurologic and extramusculoskeletal examination must be completed in the patient with suspected or known metastatic disease of the spine. The overall appearance of the patient is important, and signs of cachexia or distress should be noted. Digital rectal examination for prostate evaluation in men should be included if metastatic disease of the spine is suspected. In general, medical doctors are better equipped to evaluate the overall patient. Mindful collaboration and communication with other subspecialties is critical for providing the highest quality of patient care.

The physical examination of the spine should always include resting posture, range of motion, and palpation of the entire spine. Sites of focal tenderness to palpation and obvious deformity should be noted. Upper and lower extremities should undergo thorough sensory and motor testing of the major dermatomes and myotomes, respectively. Sensory examination includes pinprick for spinothalamic tract deficits and proprioception (great toe up or down) or vibratory sense for testing the dorsal columns. If a metastatic lesion has encroached into the epidural space and is causing spinal cord compression, there is a 65% to 80% incidence of weakness, most commonly with hip flexion. The following standardized 0 to 5 scale is used to grade muscle strength:

0—No muscle contraction
1—Muscle contracts but no movement
2—Movement in the horizontal or gravity free plane
3—Movement against gravity with no resistance
4—Movement against gravity with some resistance
5—Movement against full resistance

A thorough upper- and lower-extremity examination is extremely important. In the upper extremity, motor testing should encompass the deltoids (C5), biceps (C5-C6), wrist extensors (C6), triceps (C7), wrist flexors (C7), finger extensors (C7), grip strength (C8), and hand intrinsics with finger abduction and adduction (T1). A pronator test should be performed as well. In the pronator test, patients hold their arms extended and supinated straight out in front of them. Patients then close their eyes for 20 to 30 seconds; if an arm "drifts" into pronation and drops, the test is positive for subtle weakness from upper motor neuron dysfunction. Feigned weakness should be considered in patients who drop the arm without obligate pronation. Hyperreflexia of the upper extremities (biceps [C5], brachioradialis [C6], and triceps [C7]) and the presence of pathologic reflexes (Hoffmann's sign) are important indicators of upper motor neuron disease.

In the lower extremity, motor testing should include the hip flexors/iliopsoas (L1-L2), quadriceps (L2-L4), anterior tibialis (L4), extensor hallucis longus (L5), and gastrosoleus (S1). Again, hyperreflexia of the lower extremities (patellar tendon [L4] and Achilles tendon [S1]) and the presence of pathologic reflexes (Babinski's and clonus) are important signs of upper motor neuron disease. Digital rectal examination should be performed to determine S2-S5 sensation, sphincter tone, and volitional control. A bulbocavernosus reflex (polysynaptic, sacral mediated) is obtained by squeezing the glans penis or clitoris, and sensing an "anal wink" on rectal examination. This reflex can also be initiated with a gentle tug of the Foley catheter if one is present. Lack of a bulbocavernosus reflex can indicate spinal shock or a compressive lesion below the conus medullaris. Hyperreflexia and impairment of tandem gait (heel-to-toe walk) is suggestive of myelopathy.

PATHOANATOMY

In general, spinal metastases occur via hematogenous spread. This is thought to occur either through segmental arteries into the marrow or via retrograde flow through the extradural Batson's venous plexus, although the precise mechanism is yet to be defined.[7] Host and tumor-specific adhesion molecules and receptors are also thought to play a role in the spread of spinal metastases.[8] The pain associated with spinal metastasis is most commonly caused when the lesion within the aneural cancellous bone breaks through the surrounding cortex and contacts the periosteum, which has a high density of pain signal transmitted nerve fibers. Pain can also be related to neural compression, which presents in a more radicular pattern and is dependent on location.

Spinal metastases originate most commonly in the vertebral body followed by the pedicle, with lower vertebral levels having the highest incidence of metastatic disease (lumbar>thoracic>cervical). However, disease in the thoracic spine most commonly has associated neurologic dysfunction secondary to the closer relationship and tighter confines of the thoracic cord in the thoracic spinal column. Metastatic disease of the spine most commonly spares the intervertebral disk as it is avascular, which can help differentiate it from infection on advanced imaging. The distribution of spinal metastases is 60% thoracic, 30% lumbosacral, and 10% cervical, with prostate being more commonly seen in the lumbar spine.[9]

Interestingly, tumor cells are not directly responsible for destroying bone. Instead, tumor cells secrete various cytokines that stimulate osteoclasts (destroy bone) or osteoblasts (build bone). Thus, certain metastatic diseases are classically lytic (lung, kidney, and thyroid), blastic (prostate), or mixed (breast). It is important to realize, however, that at least 50% of bony destruction must be present to be visualized on plain radiographs. Thus, the entire clinical picture, especially the history and physical examination, remains important for an accurate and expeditious diagnostic process.

IMAGING

Advanced imaging techniques are important for characterizing the pattern (ie, blastic, lytic, or mixed) of the lesion, extent of bony destruction, and neural compression. Imaging is also used to identify other lesions that may be contributing to either pain or neurologic symptoms. As with most orthopedic entities, one should always start with plain radiographs. The most common images are anteroposterior and lateral of the affected spinal region (cervical, thoracic, and lumbar) with addition of flexion and extension lateral views to evaluate sagittal stability. Plain standing radiographs are excellent for evaluating overall spinal alignment. The classic finding for tumor on an AP radiograph of the spine is the "winking owl" sign. This can be seen as a unilateral missing pedicle profile (Image) and occurs as a lesion in the vertebral body that extends into and obliterates the pedicle. Again, small metastatic lesions are often not visualized on plain radiographs until approximately 50% of the trabecular bone is destroyed.

Advanced radiographic studies remain critical to metastatic spinal disease evaluation. Nuclear bone scan has been used in the past as a modality for detecting occult metastatic lesions. However, bone scan has been largely supplanted by modern imaging techniques, such as magnetic resonance imaging (MRI; Figure 18-1) and computed tomography (CT; Figure 18-2) scans. MRI has been shown to detect more metastatic lesions in the spine than radionuclide bone scan in various malignancies.[10] CT scanning offers improved imaging of bony detail and is commonly used in grading systems to stratify the risk of impending pathologic fracture. Computed tomography myelography can be used to assess neurocompression in patients who are unable to undergo MRI (eg, patients with a pacemaker, intracerebral clips, or implants).

Modern MRI with and without gadolinium contrast is considered the gold standard for spine imaging when metastatic disease is suspected. As seen in Figure 18-1, MRI offers excellent visualization of the neural elements (spinal cord and nerve roots) and soft tissue. MRI can also detect small lesions that are not seen with other imaging modalities, and MRI is useful in differentiating metastatic disease from other pathologies including infection. The sensitivity (92%) and specificity (94%) of MRI to detect vertebral involvement is the highest of

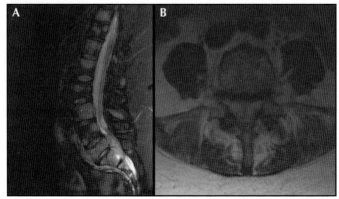

Figure 18-1. A 57-year-old patient with a history of transitional cell bladder cancer and malignant melanoma of the thumb presented to the emergency department with a 1-month history of progressive difficulty walking and profound lower-extremity weakness. Physical examination revealed weakness of ankle plantar and dorsiflexion (S1-L5), diminished perineal sensation, and poor rectal tone. (A) Midsagittal and (B) axial MRI revealed multilevel lesions with an expansile lesion at L5, resulting in complete canal occlusion.

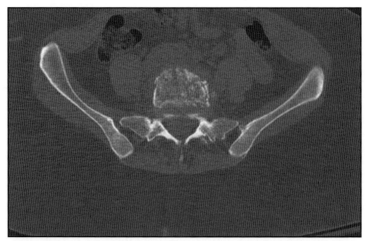

Figure 18-2. Axial CT showing a permeative lesion involving entire L5 body. (This is the same patient demonstrated in Figure 18-1.)

any imaging technique.[11] Imaging of the entire spine should be considered as multiple sites of involvement are not uncommon.

Primary tumors present with bony metastasis in 12% to 20% of cases. A metastatic work-up for an unknown primary malignancy is often required. Typically, this includes laboratory studies (ie, complete blood cell count with differential, lactate dehydrogenase, prostate specific antigen, and serum protein electrophoresis/urine protein electrophoresis [SPEP/UPEP]) and a contrasted CT scan of the chest, abdomen, and pelvis. Some practitioners recommend a whole-body technitium Tc 99m bone scan to identify additional sites of bony metastases and for staging purposes. The diagnosis is usually obvious with a known, widely metastatic primary malignancy.

In the clinical setting of a solitary spinal lesion with an unknown primary site, the differential diagnosis must be broadened to include primary bony tumor and infection. In this setting, a tissue diagnosis may be required. A CT-guided biopsy is the least invasive method to obtain a tissue sample. The sample should be sent not only to pathology but also to microbiology for the full complement of cultures (ie, aerobic, anaerobic, acid-fast bacilli, and fungus). The success rates range from up to 76% for sclerotic lesions to 93% for lytic lesions.[12]

If the location of the lesion precludes CT-guided biopsy or if the tissue obtained is insufficient, an open biopsy is appropriate. Ideally, the open biopsy should obey the general principles of tumor surgery. This includes using the same incision that one would use for planned resection, maintaining a bloodless field, and using closed drains. An intraoperative frozen section of any soft-tissue component (bone cannot be used for frozen section because it requires decalcification first) and removal of enough tissue for all requisite staining remains critical.

TREATMENT

An appropriate treatment plan must always begin with obtaining a full set of clinical data including accurate diagnosis, predicted and tumor-specific pathophysiology, life expectancy, and general medical condition. After the diagnosis is confirmed to be metastatic, neurologic stability should be evaluated by physical examination and advanced imaging as outlined above. Painful lesions without neurologic compromise or mechanical instability can usually be treated with external beam radiation therapy. Traditional external beam radiation therapy is delivered in fractionated doses over 10 to 20 sessions. This fractionation is designed to minimize injury to surrounding tissue, including the spinal cord, while delivering an effective dose to halt or reverse tumor progression. Multiple studies have demonstrated efficacy of external beam radiation therapy to improve pain and reduce associated complications, such as pathologic fractures and spinal cord compression.[13]

Although external beam radiation therapy is a mainstay of treatment for metastatic disease, there are many significant limitations. External beam radiation therapy alone is generally not an option in the presence of significant mechanical instability. In this instance, surgery is usually a more appropriate option. Some primary tumors, such as melanoma and renal cell carcinoma, are relatively radio-insensitive and may have limited response to external beam radiation therapy. Recurrent lesions often cannot be re-treated due to the risk of radiation-induced myelopathy (a form of spinal cord injury). Although studies conflict, there is some suggestion that an area that has undergone prior irradiation has a higher risk of potential wound complications if surgery is required. It is also inconvenient and difficult for patients to undergo the large number of required treatments. Patient compliance can often become an issue.

The success of traditional external beam radiation therapy and its limitations have led researchers to develop a targeted, single fraction radiation dose to the tumor bed using stereotactic radiosurgery. This technique uses multiple, mobile, narrow beam sources and a predetermined map to deliver radiation in many planes to summate at the target, thereby limiting exposure to the surrounding tissue. The machine uses internal reference points and can adjust for respirations many times per second to yield extraordinarily precise and effective results. The largest series to date demonstrated excellent short- and mid-term results, with pain and local tumor control as outcome measures.[14] There were no reported instances of neurotoxicity as well.

Some have advocated using a transpedicular, percutaneous injection of surgical cement (polymethylmethacrylate) into the vertebral body for fracture stabilization and pain relief. This technique is called vertebroplasty or kyphoplasty, with the 2 differing only slightly in method. Prior to considering

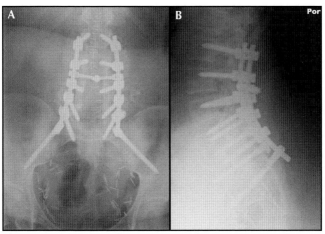

Figure 18-3. Patient underwent urgent decompression and instrumented fusion from L2 to the ilium. Postoperatively, she regained strength, bowel function, and ability to ambulate. She also underwent postoperative external beam radiation. (A) Anteroposterior view. (B) Lateral view.

this treatment, one should critically evaluate the integrity of the posterior cortex of the vertebral body as cement under pressure can extrude into the canal and result in iatrogenic neurologic injury. Tumor invasion into the epidural space is a relative contraindication to using this technique.

Surgical intervention is usually reserved for the treatment of mechanical instability, progressive neurologic dysfunction with spinal cord compression, recurrent lesions, and failure of the above treatment measures. The specifics of surgical approaches and techniques are beyond the scope of this text. Surgery commonly can achieve any or all of 3 goals: neural element decompression, mechanical stabilization, and tumor removal. Surgery is usually considered only when life expectancy is >3 to 6 months. Decompression can be achieved by removing posterior bone (laminectomy) or anterior structures (tumor/vertebral body) (Figure 18-3). Stabilization of the vertebral column can be achieved by a combination of posterior, segmental instrumentation (pedicle screws or hooks with rods), and anterior intervertebral support (bone, cage, or cement) (see Figure 18-3). The stabilizing instrumentation must provide immediate spinal stability and should not be dependent on bony healing in these circumstances.

All of the above treatments should be considered palliative measures. There may be rare instances that en bloc resection, or spondylectomy, of a solitary metastasis (renal) can be curative or significantly improve life expectancy. Treatment of a solitary renal cell carcinoma metastasis after "successful" nephrectomy is controversial. Some groups suggest that en bloc resection of the lesion (treated like a primary bone tumor) leads to improved survival rates.[15]

Multiple high-quality, Level I studies have demonstrated safety and efficacy for use of bisphosphanates, specifically zoledronic acid, to improve quality of life outcomes in patients with bony metastatic lesions. This is achieved by improving bone pain and decreasing associated complications including pathologic fractures.[16-18]

CONCLUSION

Due to the ubiquity of cancer and the high incidence of bony spread, general practitioners, orthopedic surgeons, and spine surgeons will all care for patients with metastatic disease of the spine. Most importantly, one should always have a high index of suspicion for a spinal tumor in patients with severe atraumatic back pain characterized as worse at rest and at night with accompanying cancer history or constitutional symptoms. Low-energy compression fractures without osteoporosis are also another common red flag for spinal metastasis. Work-up includes a thorough history and examination, along with complementary radiographic studies, such as CT and MRI, to fully characterize the extent of spinal involvement and associated neural element compromise.

REFERENCES

1. Constans JP, de Divitiis E, Donzelli R, Spaziante R, Meder JF, Haye C. Spinal metastases with neurological manifestations: review of 600 cases. *J Neurosurg.* 1983;59(1):111-118.

2. Aapro M, Abrahamsson PA, Body JJ, et al. Guidance on the use of bisphosphonates in solid tumours: recommendations of an international expert panel. *Ann Oncol.* 2008;19(3):420-432.

3. Coleman RE. Clinical features of metastatic bone disease and risk of skeletal morbidity. *Clin Cancer Res.* 2006;12(20 Pt 2):6243s-629s.

4. Bach F, Larsen BH, Rohde K, et al. Metastatic spinal cord compression: occurrence, symptoms, clinical presentations and prognosis in 398 patients with spinal cord compression. *Acta Neurochir (Wien).* 1990;107 (1-2):37-43.

5. Helweg-Larsen S, Sorensen PS. Symptoms and signs in metastatic spinal cord compression: a study of progression from first symptom until diagnosis in 153 patients. *Eur J Cancer.* 1994;30A(3):396-398.

6. Gabriel K, Schiff D. Metastatic spinal cord compression by solid tumors. *Semin Neurol.* 2004;24(4):375-383.

7. Yuh WT, Quets JP, Lee HJ, et al. Anatomic distribution of metastases in the vertebral body and modes of hematogenous spread. *Spine (Phila Pa 1976).* 1996;21(19):2243-2250.

8. Choong PF. The molecular basis of skeletal metastases. *Clin Orthop Relat Res.* 2003(415 Suppl):S19-S31.

9. Schiff D. Spinal cord compression. *Neurol Clin.* 2003;21(1):67-86, viii.

10. Gosfield E III, Alavi A, Kneeland B. Comparison of radionuclide bone scans and magnetic resonance imaging in detecting spinal metastases. *J Nucl Med.* 1993;34(12):2191-2198.

11. Sauvage PJ, Thivolle P, Noel JB, et al. MRI in the early diagnosis of spinal metastases of bronchial cancer [in French]. *J Radiol.* 1996;77(3):185-190.

12. Lis E, Bilsky MH, Pisinski L, et al. Percutaneous CT-guided biopsy of osseous lesion of the spine in patients with known or suspected malignancy. *AJNR Am J Neuroradiol.* 2004;25(9):1583-1588.

13. Sciubba DM, Gokaslan ZL. Diagnosis and management of metastatic spine disease. *Surg Oncol.* 2006;15(3):141-151.

14. Gersten PC, Burton SA, Ozhasoglu C, Welch WC. Radiosurgery for spinal metastases: clinical experience in 500 cases from a single institution. *Spine (Phila Pa 1976).* 2007;32(2):193-199.

15. Boriani S, Biagini R, De Iure F, et al. En bloc resections of bone tumors of the thoracolumbar spine: a preliminary report on 29 patients. *Spine (Phila Pa 1976).* 1996;21(16):1927-1931.

16. Hirsh V, Tchekmedyian NS, Rosen LS, Zheng M, Hei YJ. Clinical benefit of zoledronic acid in patients with lung cancer and other solid tumors: analysis based on history of skeletal complications. *Clin Lung Cancer.* 2004;6(3):170-174.

17. Rosen LS, Gordon D, Tchekmedyian NS, et al. Long-term efficacy and safety of zoledronic acid in the treatment of skeletal metastases in patients with nonsmall cell lung carcinoma and other solid tumors: a randomized, Phase III, double-blind, placebo-controlled trial. *Cancer.* 2004;100(12):2613-2621.

18. Saad F, Lipton A. Zoledronic acid is effective in preventing and delaying skeletal events in patients with bone metastases secondary to genitourinary cancers. *BJU Int.* 2005;96(7):964-969.

FINANCIAL DISCLOSURES

Dr. Todd J. Albert has no financial or proprietary interest in the materials presented herein.

Dr. R. Todd Allen has no financial or proprietary interest in the materials presented herein.

Dr. D. Greg Anderson has no financial or proprietary interest in the materials presented herein.

Dr. David T. Anderson has no financial or proprietary interest in the materials presented herein.

Dr. Brian T. Barlow has no financial or proprietary interest in the materials presented herein.

Dr. Mark L. Dumonski has no financial or proprietary interest in the materials presented herein.

Dr. Matthew R. Eager has no financial or proprietary interest in the materials presented herein.

Dr. Ian D. Farey is a consultant for Medtronic.

Dr. Steven R. Garfin has no financial or proprietary interest in the materials presented herein.

Dr. Greg Gebauer has no financial or proprietary interest in the materials presented herein.

Dr. Joseph P. Gjolaj has no financial or proprietary interest in the materials presented herein.

Dr. Kathryn H. Hanna has no financial or proprietary interest in the materials presented herein.

Dr. Eric B. Harris has no financial or proprietary interest in the materials presented herein.

Dr. James S. Harrop is a consultant for DePuy Spine and Ethicon. He receives grants from the Christopher Reeves Foundation North American Clinical Trials Network, and his department receives grants from Neuralstem Inc and Geron Corporation.

Dr. Alan S. Hilibrand receives royalties from Biomet for product development of an anterior cervical plate and posterior cervical instrumentation system.

Dr. Justin B. Hohl has no financial or proprietary interest in the materials presented herein.

Dr. James D. Kang has no financial or proprietary interest in the materials presented herein.

Dr. Joon Y. Lee has no financial or proprietary interest in the materials presented herein.

Dr. Joseph K. Lee has no financial or proprietary interest in the materials presented herein.

Dr. Christopher Loo has no financial or proprietary interest in the materials presented herein.

Dr. Ryan P. Ponton has no financial or proprietary interest in the materials presented herein.

Dr. Kris Radcliff has no financial or proprietary interest in the materials presented herein.

Dr. Jeffrey A. Rihn has no financial or proprietary interest in the materials presented herein.

Dr. Nelson S. Saldua has no financial or proprietary interest in the materials presented herein.

Dr. Davor D. Saravanja received travel support in 2010 from DePuy and Stryker, conducted a sales meeting presentation for Stryker Australia, and received a university fellowship.

Dr. Adam L. Shimer has no financial or proprietary interest in the materials presented herein.

Dr. Harvey E. Smith received travel support from Stryker and is a consultant for DePuy.

Dr. Brian W. Su has no financial or proprietary interest in the materials presented herein.

Dr. Ishaq Y. Syed has no financial or proprietary interest in the materials presented herein.

Dr. Jeffrey M. Tuman has no financial or proprietary interest in the materials presented herein.

Dr. Vidyadhar V. Upasani has no financial or proprietary interest in the materials presented herein.

Dr. Alexander R. Vaccaro is a consultant or independent contractor for Gerson Lehrman Group, Guidepoint Global, and Medacorp. He has received royalties from DePuy, Medtronics, Biomet Spine, Osteotech, Globus, Aesculap, and Nuvasive. He holds stock or has stock option ownership interests in Replication Medica, Globus, K-2 Medical, Paradigm Spine, Stout Medical, Spine Medical, Computational Biodynamics, Progressive Spinal Technologies, Spinology, Orthovita, Vertiflex, Small Bone Innovations, Disk Motion Technology, NeuCore, Cross Current, Syndicom, In Vivo, Flagship Surgical, Advanced Spinal Intelectual Properties, Cytonics, Bonovo Orthopaedics, Electrolux, Gamma Spine, Location Based Intelligence, FlowPharma, and R.I.S.

Dr. W. Timothy Ward has no financial or proprietary interest in the materials presented herein.

Dr. Bradley K. Weiner has no financial or proprietary interest in the materials presented herein.

INDEX

Wait...There's More!

SLACK Incorporated's Health Care Books and Journals offers a wide selection of books in the field of Orthopedics. We are dedicated to providing important works that educate, inform and improve the knowledge of our customers. Don't miss out on our other informative titles that will enhance your collection.

Throughout the *Musculoskeletal Examination Series*, you will find a thorough review of the most common pathologic conditions, techniques for diagnosis and appropriate treatment methods. These pocket-sized books include very clear photographic demonstrations, tables, sidebars, and charts, taking complex subjects and bringing them to a level that will be welcomed by all.

Series Editor: Steven B. Cohen MD

Musculoskeletal Examination of the Elbow, Wrist and Hand: Making the Complex Simple
Randall Culp MD

275 pp., Soft Cover, Due: Late 2011, ISBN 13 978-1-55642-918-7, Order# 19185, **$44.95**

Musculoskeletal Examination of the Foot and Ankle: Making the Complex Simple
Shepard R. Hurwitz, MD; Selene Parekh, MD

275 pp., Soft Cover, Due: Late 2011, ISBN 13 978-1-55642-919-4, Order #19193, **$44.95**

Musculoskeletal Examination of the Hip and Knee: Making the Complex Simple
Anil Ranawat, MD; Bryan T. Kelly, MD

480 pp., Soft Cover, 2011, ISBN 13 978-1-55642-920-0, Order #19207, **$48.95**

Musculoskeletal Examination of the Shoulder: Making the Complex Simple
Steven B. Cohen, MD

240 pp., Soft Cover, 2011, ISBN 13 978-1-55642-912-5, Order #19126, **$44.95**

Musculoskeletal Examination of the Spine: Making the Complex Simple
Jeffrey A. Rihn, MD; Eric B. Harris, MD

352 pp., Soft Cover, 2011, ISBN 13 978-1-55642-996-5, Order #19965, **$44.95**

Please visit **www.slackbooks.com** to order any of the above titles

24 Hours a Day...7 Days a Week!